Powerful Finance and Innovation Trends in a High-Risk Economy

Also by Blandine Laperche and Dimitri Uzunidis

JOHN KENNETH GALBRAITH AND THE FUTURE OF ECONOMICS (*editors*)

INNOVATION, EVOLUTION AND ECONOMIC CHANGE:
New Ideas in the Tradition of Galbraith (*co-editors with James Galbraith*)

Powerful Finance and Innovation Trends in a High-Risk Economy

Edited by
Blandine Laperche
and
Dimitri Uzunidis

Selection and editorial matter © Blandine Laperche and Dimitri Uzunidis 2008
Foreword © Luiz Carlos Bresser-Pereira 2008
Individual chapters © Contributors 2008

All rights reserved. No reproduction, copy or transmission of this publication may be made without written permission.

No paragraph of this publication may be reproduced, copied or transmitted save with written permission or in accordance with the provisions of the Copyright, Designs and Patents Act 1988, or under the terms of any licence permitting limited copying issued by the Copyright Licensing Agency, 90 Tottenham Court Road, London W1T 4LP.

Any person who does any unauthorized act in relation to this publication may be liable to criminal prosecution and civil claims for damages.

The authors have asserted their rights to be identified as the authors of this work in accordance with the Copyright, Designs and Patents Act 1988.

First published 2008 by
PALGRAVE MACMILLAN
Houndmills, Basingstoke, Hampshire RG21 6XS and
175 Fifth Avenue, New York, N.Y. 10010 Companies and representatives throughout the world

PALGRAVE MACMILLAN is the global academic imprint of the Palgrave Macmillan division of St. Martin's Press, LLC and of Palgrave Macmillan Ltd. Macmillan® is a registered trademark in the United States, United Kingdom and other countries. Palgrave is a registered trademark in the European Union and other countries.

ISBN-13: 978–0–230–55359–0   hardback
ISBN-10: 0–230–55359–1   hardback

This book is printed on paper suitable for recycling and made from fully managed and sustained forest sources. Logging, pulping and manufacturing processes are expected to conform to the environmental regulations of the country of origin.

A catalogue record for this book is available from the British Library.

Library of Congress Cataloging-in-Publication Data

Powerful finance and innovation trends in a high-risk economy / edited by Blandine Laperche and Dimitri Uzunidis.
    p. cm.
    Includes bibliographical references and index.
    ISBN 0–230–55359–1 (alk. paper)
 1. Capitalism. 2. Finance. 3. Technological innovation—Economic aspects. I. Laperche, Blandine. II. Uzunidis, Dimitri.
 HB501.P66 2008
 332—dc22
                                                              2007050168

10 9 8 7 6 5 4 3 2 1
17 16 15 14 13 12 11 10 09 08

Printed and bound in Great Britain by
CPI Antony Rowe, Chippenham and Eastbourne

Contents

List of Tables	vii
List of Figures	viii
Foreword by Luiz Carlos Bresser-Pereira	ix
Acknowledgements	xvi
Notes on the Contributors	xvii

Introduction: How does Finance Condition Innovation
Trajectories?   1
Blandine Laperche and Dimitri Uzunidis

Part I The Finance- and Knowledge-Based Economy

1. Innovation and Predation   13
   James K. Galbraith

2. Innovation, Finance and Economic Movements   19
   Dimitri Uzunidis

3. Macroeconomic Policy, Investment and Innovation   35
   Malcolm Sawyer

4. Doctrinal Roots of Short-Termism   50
   James E. Sawyer

5. Finance, State and Entrepreneurship in the
   Contemporary Economy   66
   Sophie Boutillier

6. The Political Economy of R&D in a Global Financial Context   88
   Jerry Courvisanos

7. Finance and Intellectual Property Rights as the Two Pillars of
   Capitalism Changes   110
   George Liodakis

Part II Innovation Trajectories and Profitability in Firms' Strategies

8 The Firm and its Governance over the Industry Life Cycle 131
   Jackie Krafft and Jacques-Laurent Ravix

9 Economic Change and the Organization of Industry:
   Is the Entrepreneur the Missing Link? 149
   Edouard Barreiro and Joël Thomas Ravix

10 Innovation and Profitability in the Computer Industry 167
   Christian Genthon

11 Creation and Co-evolution of Strategic Options by Firms:
   an Entrepreneurial and Managerial Approach of Flexibility
   and Resource Allocation 183
   Thierry Burger-Helmchen

12 Financial Means' Competencies and Innovation:
   Comparative Advantages between SMEs and Big Enterprises 207
   Francis Munier

13 The Financing of Innovative Activities by Banking
   Institutions: Policy Issues and Regulatory Options 224
   Elisa Ughetto

14 Innovation and the Profitability Imperative: Consequences
   on the Formation of the Firm's Knowledge Capital 248
   Blandine Laperche

Index 270

List of Tables

2.1	The cyclic evolution of the economy	24
5.1	From one paradigm to another	67
5.2	Resource potential of the founder of an enterprise: elements for a general definition	80
5.3	Big enterprises and entrepreneurs during the second half of the twentieth century	83
8.1	Governance in the early and late stages of the ILC	138
8.2	The governance of innovative firms	145
10.1	Descriptive statistics	174
10.2	Relationship between profitability and R&D intensity, 1982–89	175
10.3	Relationship between profitability and R&D intensity, 1994–2001	175
10.4	Relationship between profits and industrial organization, 1982–89	177
10.5	Relationship between profits and industrial organization, 1994–2001	177
10.6	Determinants of R&D	178
11.1	Specification of the technological and market regimes	201
12.1	Means and financial competencies	214
12.2	Distribution (in %) of the companies according to their size in the sectors by technological intensity (except energy)	215
12.3	Estimation results for 'R&D' competencies	218
12.4	Estimation results for financial competencies	219
12.5	Estimation results of 'sale of the innovation'	221
13.1	Market segmentation and relative weight of business areas by the 12 largest Italian banks (%)	231
13.2	Overview of qualitative criteria for credit risk assessment	238
13.3	Banking programmes to sustain innovative activities	240
14.1	Means of formation of the firm's knowledge capital	256

List of Figures

5.1	The socialized entrepreneur	82
8.1	Decision tree and pay-off ranking	140
8.2	Evolution of populations 1 (innovators, x-axis) and 2 (predators, y-axis)	141
11.1	The option chain in the strategic literature	185
11.2	Interaction of the option chain with the agent, firm and industry	188
11.3	(a) Real option created; (b) total number of firms during all the simulation; (c) firms active at the end of the simulation; (d) firms created; (e) firms that exited the industry	197
11.4	Box plots for 32 firms: (a) Average level of the technology; (b) maximum level of the technology; (c) number of alternative technologies; (d) average profit; (e) option-type firm profit part; (f) market share of option-type firm	200
11.5	Technological and market regimes. Regime 1 (entrepreneurial) is always on the left, regime 2 (big firms/managerial) on the right. The smallest value of each is taken to express the second on a 100 basis	204
12.1	Distribution (in %) of the companies according to the class of size	214
14.1	Knowledge capital	253

Foreword: Why Finance-Based Capitalism?

Luiz Carlos Bresser-Pereira

Contemporary capitalism is driven by innovation and finance, but these two powerful factors have different outcomes. While innovation is always growth friendly, finance has mixed consequences: it may link savings and investments in a textbook fashion, and it may bring Schumpeterian gains as it is a condition of entrepreneurial innovation. But finance may also be associated with high-risk speculation, herd behaviour and financial crisis at the macroeconomic level, and it may lead to short-term behaviour on the part of business firms that undermines rather than promotes innovation. The essays in this book deal essentially with the benefits of innovation and the risks and distortions of high-powered finance in contemporary capitalism, with an emphasis on the latter. Yet, since 2002, growth rates in all capitalist countries have been higher than ever. Innovation or technological progress transformed into reality may always be an explanation, but how do we make it consistent with the instability and financial short-termism that characterizes business behaviour in post-industrial capitalism?

In this book, the authors discuss and offer answers to these questions, but do not resort to the old saying that 'we are entering the stage of financial capitalism'. Instead, Blandine Laperche and Dimitri Uzunidis in the Introduction, most appropriately speak of 'finance-based capitalism'. Events did not bear out the prediction flowing from Hilferding's analysis of German capitalism at the beginning of the twentieth century. Finance indeed became powerful, but not for the reasons that Hilferding put forward a century ago, namely, the fusion of banking capital with industrial capital at the behest of the former, and the rise of financial capital. Even in Germany this trend did not materialize. In ownership terms, finance and production remained reasonably separate although not disconnected. Classical capitalism did not become 'financial capitalism', but became 'global capitalism' if we take into account the integration of production and finance at the world level; or 'monopoly capitalism' if we bear in mind the continuous concentration of capital by means of mergers and acquisitions; or 'post-industrial capitalism' if we consider the enormous advance of services; or 'professionals' capitalism' if we acknowledge the rise of the professional class and its association with capitalists in running the economy and sharing income, prestige, power and privilege; or 'finance-based capitalism', not because banks control productive industries but because finance is more powerful than ever and controls an increasing share of world income and wealth. Why did capitalism become finance-based if the classical explanation does not hold?

Finance has always been powerful because banks create money, and, after all, money is the objective of all economic activity. Moreover, finance is or should be powerful due to the role that it plays in financing investment, consumption and international trade; but in this case theory was never consistent with practice given the major role played by self-finance, the reinvestment of retained earnings by business enterprises: the figures in this connection continue to be surprisingly high despite the development of financial markets. Indeed, these facts and arguments fail to elucidate the overwhelming significance that finance has acquired in economic affairs since the 1980s, because they are not new, and because some of them have had negative outcomes, as in the case of self-financing for investment.

One option is to resort to more general facts that have changed the world economy and finance. The most relevant of these facts is that since the end of the gold standard money has been purely fiduciary. This major change made banks and, more broadly, the financial system more unstable, and explains why central banks are so strategic: besides controlling investment and growth by managing interest rates and exchange rates, they regulate banks and limit their capacity to create money. In this volume, James Galbraith gives an account of the increased financial power that resulted from the financial upheaval of the 1980s, brought on by the Reagan–Volcker policies that pushed nominal interest rates above 20 per cent and caused a 60 per cent appreciation of the dollar. This change, along with financial globalization or the opening of financial markets worldwide, was certainly behind the new power of finance. The brutal increase in interest rates to fight inflation and protect the dollar was an extreme attempt to protect the American economy from the market failure of stagflation. The fact that it triggered a debt crisis in developing countries and threatened a world financial crisis by depreciating the credit of the major international banks was an unintended consequence which weakened rather than strengthened the financial institutions; but those institutions eventually came out of the crisis strengthened by an awareness of the key and momentous role they perform in the global economy. The end of the gold standard and the transformation of money into a mere fiduciary asset, financial globalization, and the astonishing size and speed of international financial flows are indeed three interconnected reasons for the increased power of finance.

A fourth reason why finance has become so central and influential in contemporary capitalism is the change in the concept of capital from ownership of the means of production, that is, the net worth of the firm published in its balance sheets, to the discounted value of business enterprises' cash flow (Bresser-Pereira, 2006). Until the mid-twentieth century, business firms' valuation depended on their net worth, on the assumption that their financial statements were correct. Capital was not simply a physical quantity – a stock of buildings, machines and inventories – but its ownership. Yet since the 1960s all this has changed, as the measure of flow, not of stock, has become the basis for valuation: the value of a firm or the capital of its stockholders corresponds to the cash flow that the firm produces, discounted at a reasonable interest rate. Capital has continued to be a form of ownership, but now it is ownership of a cash flow. The basis for financing is no longer a clear sum of products whose market prices are known and reasonably stable, but rather a fluid and contingent cash flow or money flow, which depends more on managerial capacity than on fixed assets.

Fifth, we have financial innovations and the new financial agents that are involved in them. Financial innovations are many but I single out two: hedging and leveraged buyouts. Leverage appears to reduce risks but it actually increases them by expanding financial capacity. In their turn, leveraged buyouts create substantial opportunities for profit on the part of their agents, while management of business enterprises is more risky as it is under permanent threat of hostile takeover. These two innovations increase risks but the new financial agents that take advantage of them (asset management, private equity and venture capital firms) enjoy greater influence. In the past these roles were essentially played by commercial banks, but in a limited way; in the 1980s, however, they began to be performed by independent and aggressive organizations run by professionals who originally had no capital but only technical or financial skill. The role of finance was thereby amplified. Consider, first, the asset management firms. Their multiplication and competent performance had the same effect on business enterprises as the threat of leveraged buyouts: while they increased the power of finance, they accentuated 'short-termism'. More relevant, however, are the private equity firms. They play a major role in the newly powerful world of finance, but they are also the source of major market distortions. Private equity firms are so called because they were created to arrange finance for closed or 'private' business enterprises whose stocks are not quoted on the stock exchanges. But today they also deal aggressively with 'public' corporations in so far as they are responsible for initiating leveraged buyouts. *The Economist* (2007: 11) recently summarized, though it did not accept, the arguments against the behaviour of private equity firms: 'Private equity is routinely charged with all sorts of iniquity. It strips companies of assets and flips them for a fat buck. It loads them up with dangerous amounts of debt, to suck out capital for its investors. It pays scant attention to employees and suppliers. Its greedy partners avoid the tax that others have to pay.' This is not the moment to assess such criticisms. They are certainly one-sided, because, as a trade-off, private equity firms are able to push business enterprises to perform better, more efficiently, and to show higher profits. It is more relevant to consider the enormous amount of finance that these firms deal with. The value of mergers and acquisitions in the first two quarters of 2007 reached $2.7 trillion, with 465 deals worth over $1 billion each. The total was buoyed by buyouts, including the largest on record: the purchase of BCE, a Canadian telecoms company. The average size of a merger and acquisition deal was $298 million, 58 per cent higher than in the first half of 2006, and the number of hostile bids (407) was almost four times greater. Of this total of mergers and acquisitions, private equity deals already account for 35 per cent of the value in the United States and for 25 per cent worldwide. Given their combined managerial and financial abilities, private equity firms generate large returns to the rentiers and the banks that provide finance for takeovers and restructuring. Thus, private equity firms contribute substantially to the new power of finance. Self-finance may remain dominant in financing investment, but mergers and acquisitions have grown strongly since the 1970s, and the increased share of private equity firms acting as intermediaries has given new influence to finance.

A sixth and major cause of the new weight of finance in the world economic system is the likely increased participation of rentiers in the total stock of capital. I say 'likely' because I have no data to confirm or refute the hypothesis, but the logic behind the idea is clear. Given that the total stock of capital existing in the world is owned either by active entrepreneurs or by inactive stockholders, one can expect that, as the wealth of nations grows, the share of inactive stockholders tends to increase. This is true for two reasons: because there is a secular tendency for the management and the control of business enterprises to become separated, transforming stockholders into mere financial agents, and because the ageing of the population in high- and middle-income countries tends to increase pension, insurance and mutual funds' control. Since such funds are also financial organizations headed by professional managers, it is easy to understand how this enormous mass of capital owned by inactive stockholders leads to finance-based capitalism.

Thus finance-based capitalism is also professionals' capitalism – it is a form of capitalism in which financial organizations – banks, funds, private equities, asset management firms – and the bright, salaried professionals who run them concentrate power and income in their hands at the expense of inactive capitalists – the stockholders. A sign of this change is the nonconformity of *The Economist* against this fact, and the warm welcome it gives to all episodes in which stockholders regain some power. It is also the successful book *The Battle for the Soul of American Capitalism* by John C. Bogle (2005: xix) whose main purpose is to re-establish the 'true' spirit of capitalism:

> over the past century, a gradual move from *owners' capitalism* – providing the lions' share of the rewards of investment to those who put up the money and risk their own capital – has culminated in an extreme version of *managers' capitalism* – providing vastly disproportionate rewards to those whom we have trusted to manage our enterprises in the interests of their owners.

Rewards that soon turn professionals into capitalists.

Bogle believes that 'managers' capitalism is a betrayal to owners' capitalism'. It is not; it is just a new stage of capitalism in which it is transforming into professionals' or knowledge capitalism in so far as the strategic factor of production changes from capital to managerial, technical and communication knowledge. In this new stage of capitalism development, income has been becoming concentrated everywhere and especially in the United States. As Godechot (2007: 6) observes, 'mounting inequality is caused mainly by the increase in the salaries of a small elite at the top of the salaried hierarchy'. In fact, in 1970 the salaries of the top 1 per cent of the best-paid professionals represented 5.12 per cent of the total salaried mass; in 2000 this figure had risen to 12.33 per cent (Piketty and Saez, 2004).

When we start to speak about increased inequality we are beginning to discuss the negative consequences of finance-based capitalism. Some such consequences, like income concentration, are in conflict with the objective of social justice that modern societies share. Others, like 'short-termism' – a managerial distortion that forces public business enterprises that should think and act according to the medium-term approach required by investment decisions to become concerned just with what will appear in their quarterly financial statements – are obstacles to another political objective of modern democracies, namely, economic growth. Short-termism is also in conflict with moral values. The temptation to fix financial statements increases; the Enron case was just an extreme example of a practice that tends to become normal. As James Sawyer underlines in this volume, 'often short-termism in the United States takes the form of financial manipulation, even predation'. It was no accident that business enterprise fraud was the subject of John Kenneth Galbraith's last book (2004). In the times of owners' or capitalists' capitalism we had the *robber barons* as the main business entrepreneurs, now we have *individualist professionals* acting as unethically as the former. On the other hand, the increased pressure on public business enterprises to achieve short-term outcomes did not contribute to strengthening the efficient market hypothesis, as the conventional orthodoxy holds; in fact, it only distorted management decisions and the overall process of entrepreneurial innovation. Not surprisingly, an expanding economic literature, springing generally from economic schools but also from the financial departments of business schools and from law schools, shows how financial agents' endeavour to make financial markets efficient actually stimulates fraud (Williams and Findlay, 2000; Langevoort, 2002; Goshen and Parchomovsky, 2006).

Many other distortions may arise from finance-based capitalism. Among them, an old one – financial instability – is central. Neoclassical economists tried to explain it with the fiscal populism argument, but the 1997 Asian crisis showed clearly that the central cause lay elsewhere. It was the acceptance by emerging countries of insistent advice emanating from Washington and New York since the early 1990s – advice to grow with foreign savings, that is, with current account deficits financed by bonds and foreign direct investment (Bresser-Pereira, 2004; Bresser-Pereira and Gala, 2007; Gonzales, 2007).

All these distortions are discussed in this book. They are market failures that go against business innovation and growth – the central theme of the book. Yet they are not sufficiently damaging to prevent capitalism from remaining dynamic; capitalism is unjust, unstable, and not as efficient as it could be, but it is always growing and changing. But, for that it needs to be regulated and re-regulated. Since the 1970s the mantra of the neo-liberal ideology has been deregulation; but if some deregulation was necessary, most of it just increased the influence of finance, making for finance-based capitalism and spurt economic instability. In order to avoid worse consequences, re-regulation is the only option. If all markets are socially built, they are institutions that involve and require regulation; if agents in financial markets are increasingly powerful and a source of distortions and instability, they require even more regulation. But regulation of financial markets is the subject for another book. The present volume is designed to show the power and the dangers involved in high-risk finance, and the challenges that innovation poses to business enterprises and nations; and it performs these tasks very well.

References

Bogle, John C., *The Battle for the Soul of American Capitalism* (New Haven: Yale University Press, 2005).
Bresser-Pereira, Luiz Carlos, 'Professionals' Capitalism and Democracy', in Blandine Laperche, James K. Galbraith and Dimitri Uzunidis (eds), *Innovation, Evolution and Economic Change: New Ideas in the Tradition of Galbraith* (Cheltenham: Edward Elgar, 2006), pp. 17–37.
Bresser-Pereira, Luiz Carlos and Paulo Gala, 'Why Foreign Savings Fail to Cause Growth', in Eric Berr (ed.), Elgar book to be published. Available at www.bresserpereira.org.br. In Portuguese: *Revista de Economia Política* 27(1): January: 3–19.
Economist, The, 'The Trouble with Private Equity', 5 July, 2007.
Galbraith, James K., 'Innovation and Predation'. Chapter 1 in this volume.
Galbraith, John Kenneth, *The Economics of Innocent Fraud* (Boston: Houghton Mifflin Company, 2004).
Godechot, Olivier, *Working Rich* (Paris: Éditions La Découverte, 2007).
Gonzalez, Lauro, 'Crises Financeiras Recentes: Revisitando as Experiências da América Latina e da Ásia'. São Paulo: São Paulo School of Economics of Getulio Vargas Foundation, PhD dissertation, June 2007.
Goshen, Zohar and Gideon Parchomovsky, 'The Essential Role of Securities Regulation', *Duke Law Journal*, 55(4) (2006), 711–82.
Langevoort, Donald, 'Taming the Animal Spirits of the Stock Market: a Behavioral Approach to Securities Regulation'. Berkeley, Calif.: University of California, Berkeley Program in Law and Economics, Working Paper Series.
Piketty, Thomas and Emmanuel Saez, 'Income Inequality in the United States', in Anthony Atkinson and Thomas Piketty (eds), *Top Incomes over the Twentieth Century* (Oxford: OUP, 2004).
Sawyer, James, 'Doctrinal Roots of Short-Termism'. Chapter 4 in this volume.
Williams, Edward E. and M.C. Findlay, 'A Fresh Look at the Efficient Market Hypothesis: how the Intellectual History of Finance Encouraged a Real Fraud-on-the Market', *Journal of Post Keynesian Economics*, 23(2) (Winter 2000–2001), 181–99.

Acknowledgements

The chapters included in this volume were presented at the International Conference 'Knowledge, Finance and Innovation' (Second Forum the Spirit of Innovation), in September 2006, organized by the Industry and Innovation Research Unit (Lab. RII), University of Littoral Côte d'Opale (France). Many institutions supported the organization of this event and the publication of this volume. We would like especially to thank the French Ministry of Education and Research, the National Centre for Scientific Research (CNRS), the Embassy of the United States in Paris, the National Science Foundation, the French Region Nord/Pas de Calais, ISM (Institution Supérieur des Métiers, Paris), the IGS group (Institut de Gestion Sociale, Paris), the cities and communities of Dunkirk and of Boulogne-sur-Mer, the Dunkirk and Boulogne-sur-Mer Chambers of Commerce and the University of Littoral Côte d'Opale. We especially thank Edward Anthony, Frédéric Cuvillier, Michel Delebarre, Roger Durand, Claude Fournier, Roger Serre, Mark Suskin, Sylvie Vacheret and Yannick Vissouze.

We offer our deep appreciation to all the members of the scientific committee of this conference, for their help in the selection of chapters, in the construction and in the publication of this book. We especially thank Philip Arestis (University of Cambridge, UK) for his help in the publication of this book. We also thank our colleagues for their valuable help, especially Sophie Boutillier (Lab. RII, France), Jacques Kiambu (Lab. RII, France) and James Sawyer (Seattle University, USA and Lab. RII, France) who read some parts of the book and Jean-Pierre Verecken, who made important corrections to the manuscript.

Notes on the Contributors

Edouard Barreiro is Associate Researcher at CNRS–GREDEG, University of Nice Sophia Antipolis, France.

Sophie Boutillier is Associate Professor, Research Director in Economics at the Research Unit on Industry and Innovation (Lab. RII), University of Littoral Côte d'Opale, France.

Luiz Carlos Bresser-Pereira teaches Economics and Political Economy at the Getulio Vargas Foundation, São Paulo, Brazil.

Thierry Burger-Helmchen is Associate Professor of Economics at BETA, University Louis Pasteur, Strasbourg, France.

Jerry Courvisanos is Senior Lecturer in Innovation and Entrepreneurship, School of Business, University of Ballarat, Ballarat, Victoria, Australia.

James K. Galbraith holds the Lloyd M. Bentsen Jr Chair in Government/Business Relations at the Lyndon B. Johnson School of Public Affairs, the University of Texas at Austin, USA.

Christian Genthon is Associate Professor in Economics at LEPII, University Pierre Mendès France, Grenoble, France.

Jackie Krafft is CNRS researcher in economics, at CNRS–GREDEG, University of Nice Sophia Antipolis, France.

Blandine Laperche is Associate Professor, Research Director in Economics at the Research Unit on Industry and Innovation (Lab. RII), University of Littoral Côte d'Opale, France.

George Liodakis is Professor of Political Economy, Technical University of Crete, Greece.

Francis Munier is Associate Professor, Research Director in Economics at University Louis Pasteur, Strasbourg 1, France.

Jacques-Laurent Ravix is Professor of Economics at CNRS–GREDEG, University of Nice Sophia Antipolis, France.

Joël-Thomas Ravix is Professor of Economics, CNRS–GREDEG, University of Nice Sophia Antipolis, France.

James E. Sawyer is Professor of Economics, University of Seattle, USA and Research Unit on Industry and Innovation (Lab. RII), France.

Malcolm Sawyer is Professor of Economics, University of Leeds, United Kingdom.

Elisa Ughetto is Research Fellow at Politecnico di Torino, Italy.

Dimitri Uzunidis is Director of the Research Unit on Industry and Innovation (Lab. RII), University of Littoral Côte d'Opale, France and Technical University of Crete, Greece.

Introduction: How does Finance Condition Innovation Trajectories?

Blandine Laperche and Dimitri Uzunidis

In the 1970s, the large size of enterprises, considered by Schumpeter (1942) as the sign of the crumbling of capitalism's walls, was one of the symbols of the rigidity of the mass production system. The second symbol was state interventionism – government support of a production mode in crisis (Boccara, 1977) – and blamed for the economic difficulties of the period: growing unemployment and upsurge of inflation.

Two main changes have provoked radical evolutions and have opened new perspectives for solutions to economic difficulties. The first change refers to the globalization of markets – or its new phase characterized by capital market deregulation – which opened new markets to big corporations that were being stifled within national boundaries. Globalization is characterized by the predominance of deregulated finance ('big finance') over economic activities: it links markets and provides new opportunities for the development of activities (Chesnais, 2004; Stiglitz, 2006). In between free enterprise capitalism and monopoly capitalism stands finance-based capitalism. This is a new accumulation framework that is developing, addressing the economic policies of market liberalization and deregulation, such as monopoly contestability, privatization, flexible work management, integration and globalization of financial markets and instruments. As a result of these policies, the stage of the concentration of production has been replaced by the flexible organization of production. Capitalism has restructured to accommodate this flexible production mode by more flexible contractual arrangements between agents which, however, do not threaten the centralization of property and economic power. These policies have to a certain extent succeeded in depreciating the old capital. At the same time, they lead to the securitization and marketization of all individual and collective assets (science is of course part of this). The financial sphere is mobilized to promote variety and reinforce selection (Perez, 2002). The promotion of the supply of new goods, methods of production and processes and the increase of the possibilities of achieving new productive combinations are rapidly sanctioned by considerations of profitability in the 'visible' horizon. This is dangerous for sustainable growth since it exacerbates non-computable risks (Knight, 1921), as demonstrated by the Internet bubble.

The second change relates to innovation that has become the engine of firms' differentiation. Academic studies (and notably the evolutionist and institutionalist approaches) aiming to discover the secrets enclosed in the 'black boxes' of technology and knowledge (Rosenberg, 1982), have gained ground in recent years and show the importance of innovation in the economic reality. To ameliorate future risks, enterprises concentrate on the accumulation of knowledge. Knowledge is scientific and technical in order to help the firm govern market forces. It also relates to the market's structure itself (business intelligence focused on competitors, clients, suppliers; see Richardson, 1998; Penrose, 1959). The enlargement of the production scale has taken on such dimensions that from now on, the appropriation of the technological elements gathered by large companies is less costly than the raising of capital for their formation. The big firms use relations of power to create convergence centres for science and techniques, which they combine to feed their innovation process.

In these conditions, finance has become a very powerful regulatory instrument. The system works by trial and error, while finance facilitates the task. But in doing so, finance directs the applications of science to production, the financial results become a selection criterion for the research programmes and at the same time financial considerations weaken the potential for radical innovations. The current economy is thus based on finance and on knowledge. This book aims at studying the features of this finance- and knowledge-based economy. It also discusses the economic, managerial and political consequences of the contradictions between the process of short-term financial profit-making and the process of innovation which implies uncertainty, risk, long-term strategy and financing, notably in the case of radical inventions involving changes in the structures of societies.

The first part of the book studies the theoretical and doctrinal roots of the finance- and knowledge-based economy and shows the contradictions that stem from it (predation, conflict between short- and long-term behaviours). The second part more specifically, often through case studies, examines the determinants and constraints of innovation processes in finance- and knowledge-based economies.

1. The finance- and knowledge-based economy

In Chapter 1, James K. Galbraith studies American corporate decline in the current period. He describes the main phases of this decline, beginning with the Japanese challenge of the 1970s and 1980s, continuing with the financial upheaval of the 1980s resulting in a new power of finance and speculation and exacerbated by the more recent Chinese challenge. Innovation and predation are dual traits of the economic structure that has emerged since the early 1980s. We see the emergence of new entrepreneurs who bring innovation as the first trait. Alignment of CEO incentives with those of shareholders seeking rates of return leads to the development of predatory behaviour as the second trait. The main problem lies in the fact that the forces spawning innovation and those spawning predation are the same and that the market does not distinguish between innovation and predation.

This situation is, according to Dimitri Uzunidis in Chapter 2, the reason why innovations – which are at the beginning of the twenty-first century the engine of competition – do not satisfy all the social and economic promises offered by scientific and technological advancements. As a matter of fact, if we carefully study Schumpeter's approach to innovation and economic growth, we can state that the new technology, new market or new industrial organization cannot explain on their own a global economic recovery. In order to understand why technology per se fails to generate economic growth or general welfare, one must distinguish between technological potentialities (everything promised by technologies) and their real use. Consequently, we must see why, how and by whom those technologies that at present look revolutionary have been developed and how they may be socially appropriated in order to become vectors of economic growth. In the first part of his chapter, Uzunidis studies the relations between innovation and economic movements in order to show in the second part the importance of institutional and social changes for economic revival. According to the author, at the start of the twenty-first century, no large-scale economic movement will be possible as long as global solvent demand remains low and as long as finance acquires a growing surplus of value and transforms it into rent. Those aspects – role of demand and impact of finance – are studied in the following two chapters.

In Chapter 3, Malcolm Sawyer studies the relations between macroeconomic policies and innovation. He argues that macroeconomic policy, and in particular monetary policy, has been based on the classical dichotomy with a separation between the demand and supply sides of the economy. It has also been based on the assumption of a predetermined growth of supply. If one believes that, in the long run, there is no trade-off between inflation and output then there is no point in using monetary policy to target output. You only have to adhere to the view that printing money cannot raise long-run productivity growth, and therefore believe that inflation rather than output is the only sensible objective of monetary policy in the long run (King, 1997, p. 6). Malcolm Sawyer argues that when the processes of innovation and change are seriously taken into account, the classical dichotomy breaks down. Decisions made by firms with regard to investment in capital equipment and in research and development impact on the supply potential of an economy. Further demand influences investment, and also influences the rate of productivity growth, for example through learning by doing. Moreover, individuals cannot optimize with full information concerning the future when the future is inherently unknowable. The conclusion reached from the use of optimizing models with full information that there would be no inherent inadequacy of aggregate demand would not be sustained in an uncertain world. Fiscal and other macroeconomic policies are then required to address these potential problems of aggregate demand. Monetary policy also has to pay regard to the output, demand and supply implications of interest rates, and not use interest rates solely to target inflation.

The power of finance presented in all previous chapters also illustrates the impact exerted by the macroeconomic context on innovation. According to James Sawyer it induces short-termist behaviours, which are not compatible with growth-leading innovation. But what are the doctrinal roots of short-termism? In Chapter 4, James Sawyer analyses the relationship between short-term thinking and its doctrinal antecedents in neoclassical theory. More precisely he argues that a profit lacuna in the received theory creates a conventional wisdom among entrepreneurs, economics analysts, public decision makers and regulators that confuses the productivity of finance with the productivity of capital. Consequently, because of resulting insufficient oversight, self-interested, hoarding-oriented behaviours by rentiers may cause them to act in short-term, socially suboptimal ways, often gaining equivalent rewards with capitalists, even if they withhold the full menu of entrepreneurial services, particularly innovation.

In this context, what is the profile of the new entrepreneurs at the start of the twenty-first century? Sophie Boutillier explains in Chapter 5 that capitalist dynamics has changed since the 1980s. Four new factors characterize this evolution: (1) the economic liberalization and the development of financial markets; (2) the development of new technologies (information and communication technology, biotechnology); (3) the decline of public expenses; (4) the increase of mass unemployment in the 1970s and 1980s and the subsequent development of policies to boost entrepreneurship. The new entrepreneurs who emerge from this context are often presented as revolutionary entrepreneurs, just like those of the nineteenth century who were at the forefront of the industrial revolution. However, S. Boutillier considers here that the new entrepreneurs are neither revolutionary nor heroic. The current period is seeing the emergence of the 'socialized' entrepreneur. This entrepreneur may, as a matter of fact, be the result of a big firm's strategy (improvement of its flexibility, its capacity to innovate), of the state's policy (economic growth and reduction of public expenditure) and of the speculative behaviour of financiers.

The next two chapters examine in more detail how this context of a finance-based economy influences the process of innovation. In Chapter 6, Jerry Courvisanos focuses on the research and development (R&D) process. He develops a political economy approach based on a dilemma created by the dual role of R&D. On the one hand, R&D aims to generate knowledge and promote the development of new ideas into products, processes and services (called 'innovations') that drive economic growth. On the other hand, it entrenches monopoly power within corporations that undertake R&D and then patent or copyright the results. The former role encourages the extension of the frontiers into new knowledge, while the latter tries to limit knowledge by marginal incremental marketing-based improvements that have more to do with planned obsolescence. The medium-to-large firms who can finance their R&D chiefly through earnings retained out of previous profits, have accountants and risk-averse managers who tend to strongly support R&D funding based on the short-term view. The small young start-up R&D firms have no substantial internal funding for R&D, and depend on public funding for support, but they have a poor track record as the vast majority of these firms fail. This dilemma in R&D is evaluated in the context of globalization of the financial system, in order to identify possible solutions.

In Chapter 7, George Liodakis develops the following argument: his view is that the current restructuring of capitalism shows the role played by finance and intellectual property in the innovation process. He studies the evolution of the sources of financing of R&D and technological innovation, and underlines the growing role of private financing. Because it is necessary to be profitable in the short term (what we can call the 'profitability imperative'), private financing leads to the expansion of some branches of production (those which can be subject to commercial exploitation) at the expense of public goods production, and to an increasing concern of corporate capital for the appropriability of innovations stemming from R&D activities. The author studies the impacts of these new trends and clearly shows their limits for the rate, content and direction of technological development.

The current economic context is largely dominated by finance, and as developed in many of the chapters of this first part of the book, by short-term behaviour, speculation and predation. However, at the same time, the economic context is also characterized as a knowledge-based economy, where new technological swarms (information and communication technology, biotechnology) give the opportunity to disseminate new productive combinations. But what are the determinants of innovation in this finance- and knowledge-based economy? This is the issue dealt with by the authors in the second part of the book.

2. Innovation trajectories and profitability in firms' strategies

In Chapter 8, Jackie Krafft and Jacques-Laurent Ravix ask the following question: what would be a good mode of governance in order to boost innovation in firms? So far, the predominant thesis has been that there should be a superior model promoting optimality disclosure of information and transparency. But this thesis is greatly contested, since the adoption of a unique and universal set of rules and arrangements neglects the diversity and heterogeneity of firms, industries, as well as institutional contexts. In this chapter the authors connect the literature on the industry life cycle with the literature on the governance of large and small enterprises. They show, as an intermediate result, that the governance of small, young and innovative firms in the early stages of life should benefit from a mode of governance based on cooperation and assistance to stimulate innovation, while a large mature and routinized enterprise should impose a mode of governance based on control of the manager's actions in the interests of shareholders. However, the latter mode of governance favours short-term choices that may be detrimental to the development of innovation. Krafft and Ravix thus advance the idea that new principles of governance should be proposed for innovative corporations (large or small), by defining the notion of 'corporate entrepreneurship' within which managers and investors are collectively involved in the coherence and development of small, but also large, innovative firms.

Moreover, is innovation the prerogative of the entrepreneur and thus of the small enterprise? According to Edouard Barreiro and Joël Thomas Ravix in Chapter 9, we can distinguish two types of entrepreneur in the economic literature. The first is inherited from the Austrian tradition: seen in this perspective, the entrepreneur is the agent of novelty and change. The second type appears in the theory of the firm. In this case, the entrepreneur is a virtual agent who exists but has no real functions in the work of the economic system. For example, we may emphasize that although Ronald Coase evoked this agent in his framework, the system reaches a spontaneous balance without any human intervention. In this chapter, the authors propose to fill the empty space between the analysis of economic change and the theory of the firm and consider the entrepreneur in a Marshallian perspective. This approach, introduced by Marshall and developed by Penrose, enables one to link innovation and economic development to the organization of the industry. Hence, it becomes possible to study the firm and the change in a single framework.

Economic change may thus be the result of the small enterprise or of the big firm's activity. But what is the nature of the relation between innovation and profitability? Does an increase in R&D expenses automatically lead to an increase in profits? Chapter 10, written by Christian Genthon, explores the nature of the relationships between innovation and profitability, by analysing the computer industry over an extended period of time (20 years, from 1982 to 2001). The aim is to identify the determinants of innovation taking this particular industry as an example. After presenting the computer industry dynamics over a long period of time and analysing the literature focused on the relationship between innovation and profitability, Genthon proposes an empirical study using a proprietary database built around the top 60 firms in the sector. It demonstrates that there is no relationship between R&D spending and profitability. The determinants of profits depend mainly on growth but also on the period studied and therefore on the industrial organization of the industry. Moreover, the determinants of profits and innovation are not only 'industry specific', but also 'historically industry specific'.

If we study the behaviour of enterprises according to a firm's size and power, however, we can assess the role played by different kinds of capacities gathered or built by enterprises in the innovation process. In Chapter 11, Francis Munier analyses the relation between the size of the firm and innovation based on the concept of competence to innovate. This approach takes into account a large variety of innovative behaviours, primarily those prevailing in small and medium-sized enterprises. The author analyses empirically the probability of having competencies according to three classes of size, in order to point out the comparative advantages between SMEs and big enterprises. Munier distinguishes financial and means competencies (human capacities for example) which are also relevant to enterprises. The inquiry made into French industrial firms empirically verifies the Schumpeterian conjecture according to which big companies would be more able to implement a research project in order to innovate.

To understand innovation at the microeconomic level, it is important to understand the decision-making process within the enterprise. In particular, how do firms succeed in specific market and technology regimes, to allocate limited research resources between alternative technologies offering uncertain prospects? This question is all the more important when profitability becomes a major objective. In Chapter 12, Thierry Burger-Helmchen compares different approaches. The first follows the evolutionary representation of decision-making processes under bounded rationality. The second approach based on real option reasoning encourages flexibility by creating and nurturing a portfolio of technologies and postponing the decision to specialize. Burger-Helmchen explores the links between the characteristics of the firm, its technological options and the evolution of the industry. To achieve this, he describes the contrasting theoretical assumptions behind each of these approaches and builds a simulation model to compare their implications. The simulation highlights the dualities and complementarities between the entrepreneur and the manager in the firm.

Innovating at the microeconomic level requires the availability of financial resources. Schumpeter has explained well the hard task of the entrepreneur to convince the banker. In the finance- and knowledge-based economy, new financial techniques have been developed (e.g. the now well-known venture capital mainly used to start innovative businesses). At the same time, however, credit support to R&D activity should be used, since new norms have been developed to facilitate this bank intermediation. In Chapter 13, Elisa Ughetto investigates to what extent the convergence of banks over risk-adjusted capital standards set by the new Basel Capital Accord may affect the way in which they screen innovative firms. She also gives an overview of the existing forms of credit support to R&D activities. The study is built upon a survey on 12 main Italian banks. The survey provides interesting insights into the use of non-financial parameters to assess the creditworthiness of potential borrowers and into the architecture of internal rating systems in the light of Basel II requirements. Results suggest that the majority of banks do not consider intangibles as meaningful determinants in credit risk assessment. This could imply that the sole implementation of the Accord might not lead to reduction of informational asymmetries between lenders and borrowers as might be expected. However, such an effect could be compensated by specific measures provided by single financial intermediates.

Finally, in Chapter 14, Blandine Laperche studies the impact of the profitability imperative that firms have to face in the knowledge- and finance-based economy on the way firms manage their innovation processes. Currently, the formation and the protection of a knowledge base, called here 'knowledge capital', is of major importance to firms, since it will be at the origin of the firm's innovative capacity and thus of its power in global markets. Laperche develops the idea according to which there is a growing contradiction between the socialization of knowledge capital (which means that many institutions are involved in its formation) and the tendency towards its oligopolistic appropriation, the latter being made easier notably by the evolution of patent laws and practices. This contradiction is explained by today's profitability imperative and its related necessity to reduce the risk for, the cost and the length of technological progress. It results in a growing danger hanging over scientific commons, which raises the issue of the disequilibrium between public and private activities, leading to 'private opulence and public squalor' (Galbraith, 1958). This means that all non-competitive activities in the short run could be neglected by private enterprise, which could have negative impacts on the future of our societies. This is an important message given by the following chapters of this book.

References

Boccara, P., *Études sur le capitalisme monopoliste d'État, sa crise et son issue* (Paris: Editions sociales, 1977).
Chesnais, F., *La finance mondialisée* (Paris: La Découverte, 2004).
Galbraith, J.K., *The Affluent Society* (Boston: Houghton Mifflin, 1958).
King, M., Lecture given at the London School of Economics (1997).
Knight, F.H., *Risk, Uncertainty and Profit* (1921) (London: Reprints of the London School of Economics and Political Science, 1948).
Penrose, E., *The Theory of the Growth of the Firm* (1959) (Oxford: Oxford University Press, 1995).
Perez, C., *Technological Revolutions and Financial Capital. The Dynamics of Bubbles and Golden Age* (Cheltenham: Edward Elgar, 2002).
Richardson, G.B., *The Economics of Imperfect Knowledge: Collected Papers of G.B. Richardson* (Cheltenham: Edward Elgar Publishing, 1998).

Rosenberg, N., *Inside the Black Box: Technology and Economics* (Cambridge and New York: Cambridge University Press, 1982).
Schumpeter, J.A., *Capitalism, Socialism and Democracy* (New York: Harper and Brothers, 1942).
Stiglitz, J., *Making Globalization Work* (New York: W. W. Norton, 2006).

Part I
The Finance- and Knowledge-Based Economy

1
Innovation and Predation
James K. Galbraith

In 1942 Joseph Schumpeter published *Capitalism, Socialism and Democracy*, a long book now mostly forgotten but with one short and immortal chapter dealing with technical change. In the phrase 'gales of creative destruction' Schumpeter called attention to the Janus face of innovation: its capacity to enrich and to destroy. On one side, innovation brings down costs: the supreme achievement of capitalism lay not in making more silk stockings for kings but bringing them within the reach of shop girls in return for steadily decreasing increments of effort. On the other side – facing producers instead of consumers, if you will – innovation is the battering ram that brings all the Chinese walls crashing down. Moreover, since technical change required the power and the resources of larger organizations, it justified the existence and even the conduct of large business firms. What appeared to be monopolistic abuse of power in a static view was transformed into the engine of productivity itself, once viewed in temporal and dynamic perspective.

I mention Schumpeter because while his name is remembered his analysis was mostly discarded.

Neoclassical growth theory took a very different point of view. Determined to preserve the infinitely many, infinitesimally small, zero-profit image of the competitive firm, yet confronted with the overwhelming importance of the Solow residual, neoclassical theory had to ask itself, where does progress come from? And since it would hardly be generated by a zero-profit enterprise, the answer had to be: from God. Disembodied, Hicks-neutral technical change was invented, so to speak, to prevent Schumpeter's Janus face from tormenting the minds of graduate students, and so to rescue from oblivion the owner-entrepreneur and the small business firm. For God, substitute 'the research lab' and you have the map that some economists and many others still carry in their heads.

Nevertheless, as theory developed in the 1950s and 1960s, a handful of technical economists continued to wrestle with aspects of Schumpeter's central thesis. These included:

- Evsey Domar (1946), whose contributions to the theory of growth included an analysis of scrapping – the destructive phase of creative distribution.
- W.E. Salter (1960), who worked on the problem of capital vintages, introducing a time dimension of the productivity of the capital stock.
- Nicholas Kaldor (Kaldor and Mirrlees, 1962), who suggested a technical progress function as an element of macroeconomic and growth theory.
- Andreas Papandreou (1949), whose concept of circular heteropoly tackled the key taxonomic issue in the definition of monopoly in industrial organization, and providing a consistent distinction between Schumpeterian innovators and the backwater of owner-entrepreneurial competition.
- Later on, Richard Nelson and Sidney Winter (1982) would attempt with some success to blend some of these ideas into formal models of evolutionary process in the theory of the firm.

Forgive me for arguing that these efforts fall short in failing to peer deeply inside the firm, in focusing on the vapour trail as though the electron making it could not be directly observed, when in fact it can be, and therefore in failing to consider seriously enough the role of *organization* in the conduct of the large business firm. In this respect the crucial contribution belongs to Galbraith and the central work is his *The New Industrial State*, published in 1967.

In *TNIS* Galbraith explained the interior structure of organizations built to manage complex technological process. He argued that once power passes within the organization it passes completely – the technostructure takes command. The CEO, celebrated as the innovator/entrepreneur, is in fact a figurehead who does not and cannot know very much. Shareholders and the stock market generally seemed to have lost relevance, except when as sometimes happens the corporation got into trouble. The consumer became not the driver but merely another adjunct of the planning process – to be manipulated so far as possible. Firms of course coordinated with other firms – implicitly on prices and critically so in the expansion of plant, so as to guard against surplus capacity and the outbreak of destructive competition. Innovation was thus to John Kenneth Galbraith a process necessarily managed for well or ill by the large business organization.

Obviously the image of large corporations successfully and smoothly managing the innovative process took a beating in the 1970s and 1980s. What happened then? Many rejected Galbraith's model of corporate power, concluding that the world really was classically competitive after all. In the 1990s we were told that the age of the owner-entrepreneur had returned. So perhaps Galbraith and also Schumpeter had been wrong. Perhaps the innovative process could really be understood within a neoclassical competitive frame. Perhaps technical progress really did fall from the heavens after all. In thinking about innovation policy, obviously it matters which model you choose.

I wish to draw a different picture of this fate of the great corporation – one that builds on Galbraith's analytical vision rather that attempting to refute it, for the major phases of this decline are strongly illuminated by the beacons of *TNIS*, even though the specific historical details were not and could not have been foreseen in that book.

To begin, the Japanese challenge of the 1970s and 1980s, especially in steel, motor vehicles and electronics, is best seen as the incursion of one planning system on to the turf of another. It was handled even by Reagan's free market ideologies, politically – with a carefully negotiated compromise that enabled the Japanese to earn high profits even while limiting the rate of market penetration. The resulting defeat for the American firms was bad but not nearly as bad as an unmanaged market would have been.

The next major phase in American corporate decline was the financial upheaval of the 1980s, brought on by Reagan and Volcker and 20 per cent interest rates, the 60 per cent overvalued dollar and the attack on countervailing power. The corporate collateral damage from this effort to restore a world of owner-entrepreneurs, dominated by bankers, was enormous. In the 1960s, my father had simply not imagined any government would be so extreme as to try.

A corollary of the new power of finance and the destruction of unions in 1980 was a search for higher rates of return and the use of corporate restructuring as a tactic in that search. Eventually what proved effective was the breaking up of information technology into independent firms, independently capitalized in the financial markets. Thus the fate of the technology sector, previously tied to the needs of the using corporation, became tied instead to the ebb and flow and occasional speculative fevers of the financial sector.

At this point there emerged the myth of the newly resurgent owner/entrepreneur. But I think close examination of such New Economy firms as Intel and Microsoft shows otherwise. Robert Noyce – a genuine scientist-businessman, sold to IBM and to the government when he built Fairchild Semiconductor and mainly to other companies when he built Intel. He was technostructure personified, a committee man, largely unknown to the public. Bill Gates needed a different public profile because Microsoft sold to the public – but Gates was never the scientific leader of Microsoft. Neither Gates nor Noyce resembled the classic owner/entrepreneur.

Galbraith supposed that the technostructure would for institutional reasons keep CEO compensation under control and that failure to do so would prove organizationally disastrous. He was wrong only in the predicate. For the financial *coup d'état* of the 1980s led to the alignment of CEO incentives with those of shareholders seeking rates of return that ordinary business methods could not deliver. And this led perfectly predictably not to institutional reforms which were anyway beyond the CEO's power but to a class of executives bent on the predatory destruction of the organizations they controlled, in the name of maximizing shareholders' value.

In extreme but not uncommon cases the predatory methods took on the coloration of 'control fraud' – the systematic looting of corporations by CEOs or those under their direct protection. Two features of this phenomenon especially stand out:

1. The fact that firms in this situation do not generally evolve into criminal enterprises because of behavioural incentives – the so-called moral hazard explanation – but rather are taken over by criminal networks for this precise purpose, and
2. The fact that the markets invariably reward predatory behaviour, which generates high but fraudulent performance reports, with high stock market valuations – thus punishing all firms who do not engage in predatory conduct.

It is obvious, I think, that the forces spawning Intel or Cisco and the forces spawning Enron or World Com – the forces spawning innovation and those spawning predation are *the same*: the autonomous power of financial markets to reward *whom they choose*. But the inability of the financial markets to distinguish Noyce from Ken Lay or Dennis Kozlowski, Intel from Enron or Tyco, is a sign of a *major* problem. Innovation we must encourage, predation we must control and preferably not only after the fact. Yet the 'market' is obviously incompetent in this task.

The fourth and final source of the great American corporate decline much noted nowadays is of course the Chinese challenge; namely the capacity of that country's firms and workforce to supply world markets with seemingly unlimited quantities of high-quality goods, taking market share seemingly at will. This is commonly attributed to low Chinese wages, but the explanation fails, as anyone who goes there can see; Chinese real wages are not low by Third World standards, and anyway one cannot extract First World quality from slave labour.

The resolution of the China puzzle lies, I believe, in noting the special character of the Chinese manufacturing firm. Because the capital market is not all-powerful, the firm is not at risk of closure through mergers and acquisitions. Because the banks will cover losses, it does not have to earn a positive rate of return. Because it can draw on Western design and technology, it does not have a large technical overhead or advertising budget. So it can, and does, concentrate on improving practice, reducing costs, moving down the learning curve; it does not adjust production to sales, but dumps the excess on the home market, where low and falling prices are the counterpart of rising real wages. The Chinese firm is, in short, the modern epitome of Galbraithianism: a principle and its success reflect the competitive superiority of a Galbraithian system.

In sum, innovation and predation are dual traits of the economic structure that has emerged since the early 1980s. It is a structure that moreover concentrates the benefits of innovation on a small class – the IT entrepreneurs – who were at the heart of rising US inequality in the 1990s, and by weakening the formerly great corporation and subjecting it to financial demands that ordinary business practice could not meet, the system has fostered a culture of looting not different in essentials from that which destroyed the industrial firms of the socialist countries in the early 1990s.

We are of course much richer and have further to fall. But the competitive disadvantage of this system will, I think, fairly ensure that these problems must be dealt with, and we ought to be thinking about new institutional visions that might preserve innovative capacity while getting predatory conduct under control.

Bibliography

Domar, E., 'Capital Expansion, Rate of Growth and Employment', *Econometrica*, 14 (2) (1946), 137–47.
Galbraith, J. K., *The New Industrial State* (Boston, Mass.: Houghton Mifflin, 1967).
Kaldor, N. and J. A. Mirrlees, 'A New Model of Economic Growth', *Review of Economic Studies*, 29 (1962), 174–92.
Nelson, R. R., *The Sources of Economic Growth* (Cambridge, Mass.: Harvard University Press, 1996).

Nelson, R. R. and S. Winter, *An Evolutionary Theory of Economic Change* (London: The Belknap Press of Harvard University, 1982).

Papandreou, A. G., 'Market Structure and Monopoly Power', *American Economic Review*, XXXIX (1949), 883–97.

Salter, W., *Productivity and Technical Change* (Cambridge: Cambridge University Press, 1960).

Schumpeter, J. A., *Capitalism, Socialism and Democracy*, 3rd edn (New York: Harper & Row (1942), 1962).

2
Innovation, Finance and Economic Movements
Dimitri Uzunidis

According to the Austrian economist J.A. Schumpeter (1883–1950), capitalism is a system that cannot tolerate immobilism, routine; capitalism is not and will never become stationary. Since the eighteenth century and for numerous economists, innovation has lain at the origin of economic growth and social welfare: new technologies, new labour organizational structures, new consumer goods, new communication tools, new needs. Periods of decline follow after periods of economic wealth. Schumpeter established a consistent link (Schumpeter, 1939) between novelty and economic growth. His theory about economic cycles is a source of inspiration for all those economists and politicians who consider that technological progress is the solution to all the evils of capitalism: unemployment, social inequalities, inflation. The watchword is everywhere the same: energies must be freed in order to design new means of enrichment so that, in that context of restructuring, all the social categories will benefit from the situation. For example, information and communication technologies or 'genetically modified organisms (GMOs)' may be turned into powerful mechanisms of economic growth. Information and communication technologies encompassing both computers, new worldwide telematic networks (such as the Internet) and industrial automatons have replaced painful and tedious jobs and created new industrial and commercial activities and services. GMOs boost agricultural yields – and offer new perspectives in the medical field for the greatest benefit of the most deprived citizens. It is often proclaimed that the information and genetic revolution is the base of a new model of capitalism and celebrates the opening of a new era of economic wealth.

If one looks more closely at Schumpeter's approach and at the ideas proposed by his followers, we cannot be really optimistic about the growth opportunities resulting exclusively from newly designed technologies. In order to understand why technology per se fails to generate economic growth or general welfare, one must distinguish between technological potentialities (everything promised by technologies) and their real use. Consequently, we must see why, how and by whom those technologies that look presently revolutionary have been developed and how they may be socially appropriated in order to become vectors of economic growth. In the first section, we propose to discuss the relation existing between innovation and economic cycles (or movements) and to study, in the second section, the importance of the institutional and social changes that are indispensable for economic revival.

2.1 Capitalism in movement

2.1.1 The foundations of innovation

Within capitalism, science is viewed as a stock of knowledge-nourishing techniques (see Nef, 1953). Science constitutes a pool of productive forces since production is merely a technological application of science (Marx, 1977, II, p. 220). Habermas (1973, p. 43) considers that following the emergence of large-scale industrial research, science, technology and innovation have been integrated into a single system. Capitalism provides a framework for the systematic application of science to production which, in its turn, stimulates the development of scientific knowledge related to the laws of nature and the universe. Aiming at production, capitalism reorientates the pool of scientific and technological knowledge thus accumulated. Hence, science becomes a productive force at the service of the enterprise. Capital contributes to give a scientific form to production (Marx, II, 1977, p. 187). Since the nineteenth century and according to L. Karpik (1972), science is no longer exclusively related to the discovery of hidden relations existing between specific phenomena, but is also the promotion of a natural order so that it becomes the very foundation of industrial activity. Consequently, economic relations determine the emergence of new technologies, goods, services or markets, generate norms and create a framework for the accumulation of capital. In this respect, economic relations govern the application of scientific knowledge and determine the scope of scientific research. Resulting technologies and innovations come from the transformation of knowledge into competencies that are necessary for production and accumulation. The technological use of science is the main factor in the development of the means of production: the latter is an indicator revealing to what extent universal social knowledge has become a direct productive force. Starting from the nineteenth century, the increasing size of enterprises and the volume of capital they detain or mobilize, have led to the integration of science into immediate production: the resulting innovations contribute to the size increase and commercial strength of the modern firm.

Innovation defined by J. A. Schumpeter as being a new 'combination of productive resources' corresponds to the process of generation and private appropriation of a whole array of resources (scientific, technological and financial) which, whenever combined by the enterprise or a group of firms, gives birth to new technologies and goods and promotes the opening of new markets. The designing of new goods is a major element of innovation. Here, big enterprises with abundant capital are at a considerable advantage. They are capable of financing research teams and experiment with a considerable number of innovations in the hope that one of them will finally emerge as being the best (Robinson, 1977). Supply creates its own demand due to the insight and pugnacity of the entrepreneur, relayed by the big company.

Science becomes a capital asset under the pressure of competition and possible political and social hazards. The authority and market power of a given enterprise depend on its capacity to master technological packages with a view to subsequently innovate. Innovation strengthens the competitive position of the enterprise and enables it to face the uncertainties of the market: running out of demand, increase in the price of productive resources, entry of new competitors, social unrest, stringent rules, etc. Therefore, science is more and more used for the design, development and distribution of innovative technologies and new goods and services. The obstacles erected against the accumulation and intensity of competition determine the speed with which capital and economic activities are being renewed. Therefore, these obstacles determine the modalities of the integration of science into production and cannot be isolated from the global development of productive forces.

Competition requires the continuous extension of capital and imposes the immanent laws of capitalistic production in the form of external regulations determining the behaviour of every individual capitalist (Marx, 1976, I, p. 241). Thus, in order to avoid disappearing, the enterprise must simultaneously innovate and grow. Innovation leads to the depreciation of old assets whose profitability is declining and creates a set of conditions favourable to a new investment wave. The more the rhythm of capital renewal accelerates, the more the marginal cost of capital increases and the more financial needs escalate. The intervention of finance facilitates the merging of capital (centralization) leading to the formation of big enterprises, so that capital and its expansion may be considered as both the starting and endpoint, the motive and aim of production. Consequently, the economy has a tendency to develop productive forces as if their sole limit were the absolute power exerted by society. Nevertheless, this tendency is in permanent conflict with the purpose, the investment of existing capital. Recurrent crises lead to the lethargy or destruction of a certain type of productive forces. After that destruction, the revival of the accumulation process will not be possible without a deep alteration of the foundations and norms of accumulation (new social organization of labour, new competition rules, new technologies, new management institutional mechanism and new modes of economic regulation).

2.1.2 Crisis and renewal

Following this argument, technological innovation (introduction on the market of new goods and services and new productive tools and methods) plays a central role in the explanation of economic cycles or, even more important, of economic movements. In the economic theory, cycles are regular periods (growth, peak and decline) of economic activity. This mechanical analysis is, however, questioned when 'exogenous factors' (as for example social struggles, war, speculation) are taken into account. In fact, Schumpeter considers that innovations increase the demand for means of production, reduce the production costs resulting from the introduction of labour-saving machines and eventually reduce the cost of material assets through, for example, the augmentation and diversification of the volumes of goods produced. Any increase in the capacity of the offer generates a growth in demand due to new consumer needs, the possibility given to consumers, entrepreneurs and wage-earners to anticipate and to resort to credit. The combined augmentation of investments, distributed income and profit will place the economy in conditions favourable to growth. This prosperity phase that follows the stimulating initial impact of radical innovations (also called 'major') will be extended by minor, incremental innovations (namely, marginal or even fundamental improvements and alterations in consumer goods, production goods and methods resulting from the major upheaval observed in the economy and markets triggered by the commercialization of radically new technologies). For example, the motor vehicle was viewed as a radical innovation until it became a standard in the transportation field, adopted implicitly (consumption as a sign indicating the belonging to a specific social group) or explicitly (consumption based on the use made of the consumed good) by the major part of the population. The enlargement of its range as well as the development of engines, materials, safety or comfort are only minor innovations which, due to social pressure (fashion, use, cost) or the strategic and competitive initiative of entrepreneurs (diversification, cost reduction) had for sole consequence the renewal of the supply in the limits prescribed by the consumption norm. The application of new automatic technologies combines well with a flexible management of the labour force (unstable jobs, lengthening of the annual work time, multiplication of teams, etc.) in order to increase the relative and absolute value created and acquired by the enterprise. This part of the increasing surplus value generates higher profits for strongly concentrated enterprises even if selling prices stagnate or fall in relative terms.

Similar to the conquest of new foreign or domestic markets, the renewal of supply illustrates the reaction of enterprises confronted with competition: the purpose is to enlarge or keep their market and guarantee their sources of profit, even more since investments injected in scientific research and new product development or intended to renew the existing supply are combined with investments made in order to streamline the production process and increase productivity. Redundancies are more and more frequent, the purchasing power is constantly shrinking and enterprises are fighting on markets that have an international character but also restricted growth potentialities. Compared to production capacities and to the cost of investments made, demand looks more and more insufficient. Growth is losing pace. Excess of investments, explained by the exaggerated optimism observed during the years of strong growth, only partly explain the turnaround of the cycle. Losses, bankruptcies, layoffs, the economy is entering a period of crisis that will last until surplus capital (unused production capacities, excessive indebtedness) is absorbed or disappears. Once more the effect of 'vacuum cleaning' liberates entrepreneurship so that a new wave of major innovations will be able to trigger a new cycle. This process of 'creative destruction' explains the cyclic evolution of the economy. In this case, crisis cannot be separated from growth: crisis is in itself a healthy factor because it facilitates the revival of the economy and the elevation of the standard of general welfare. By 'creative destruction', we mean the combined norm of consumption and the social symbols associated with that standard. If we refer again to the example of the motor vehicle or electrical appliances and although if it is true that a million individuals worldwide have no possibility of access to these commodities or are even in a position to acquire them, they do it at the cost of considerable indebtedness and hardship. For the major part of the population, the possession of these categories of goods becomes first a sign of social success, then of economic welfare. On the other hand, lifelong learning, the prevention

Table 2.1 The cyclic evolution of the economy

	Prosperity	Recession	Depression	Recovery
'Industrial revolution cycle' (textiles, iron and steel, steam)	1787–1800	1801–1813	1814–1827	1828–1842
'Bourgeois' cycle (railways)	1845–1857	1858–1869	1870–1885	1886–1897
'Mercantilist' cycle (motor vehicles, electrics, chemistry)	1898–1911	1912–1925	1926–1939	(1940–1950)
'Interventionist' cycle (oil, plastics, electrical engine)	(1951–1965)	(1966–1975)	(1976–1985)	(1985–1995)
'Managerial' cycle (microelectronics, materials, biotechnology)	(1996–)	?	?	?

Note: The first cycles were defined by J.A. Schumpeter in *Business Cycles* (1939). The interventionist and managerial cycles are our updating of Schumpeter's analysis.

of diseases or environmental protection merely attract the attention of the actors in the system when they are successfully transformed into divisible goods and services that can be appropriated individually.

If we follow and extend Schumpeter's approach (Schumpeter died in 1950) and also refer to the economist and mathematician N. Kondratiev (1935), we may identify five economic cycles or to be more precise, five long accumulation movements accompanying five waves of radical innovation: textile, iron and steel industry, steam at the end of the eighteenth century; railways in the middle of the nineteenth; electricity, car industry, chemistry at the beginning of the twentieth; oil, plastics, electrical engines in the middle of the twentieth; microelectronic, materials, biotechnologies to the end of the twentieth century (see Table 2.1). The cyclic evolution of the economy proposed by Schumpeter and the trajectory described by those economists who find their inspiration in his works are not mechanistic. The major transformations of the market cannot be separated from the deep changes that occur in the social relations related to production and ownership (that is why some economists prefer the word movement to the one of cycle). Therefore, Schumpeter validated the conclusions reached by Marx when the latter says that starting from a specific point, capitalistic production relations become socially inappropriate in the face of the extraordinary development of productive forces and the socialization of the creation and circulation of values.

2.2 Innovations and long-lasting accumulation movements

2.2.1 Innovation and institutional changes

Economic activity is permanently questioned and renewed by innovations. However, innovations do not follow a regular progression: their introduction follows a discontinuous process. The importance of innovation is measured by the cumulative external effects it provokes at the level of the economy: cost reductions in all the production activities, labour productivity increases, possible compatibility with the specific technologies of each industrial sector, diffusion to tertiary activities. For example, since the end of the 1960s, microelectronics have permeated all spheres of economic life. The microprocessor is considered as the core technology that made it possible to open up new markets and new accumulation domains, either through the creation of new goods and services (PCs, Internet, network services, multimedia) or through the redefinition of already existing electromechanical processes (machine tools; systems used to transport, treat and store information). Combined with other techniques, that core technology leads to the emergence of successive innovation clusters provoking imbalances wherever they appear: capital abandons inert and saturated sectors to migrate to sectors characterized by a high degree of growth and profitability. The logic of innovation clusters is linked to that of the organization of the enterprise according to the same cluster principle.

As a matter of fact, the migration of productive and financial capital towards the achievement of new combinations of means of production and the launching of new commodities are at the origin of the constitution of collectives of workers that are always more complex, costly and incorporate an increasing number of wage earners, firms and institutions. The socialization of the creation of wealth in the capitalist framework unceasingly increases; markets are transformed and extended; production requires increasing volumes of capital as well as a sophisticated labour force and new rules applicable to the organization of production. According to J. K. Galbraith (1967), innovation and, in particular, technology go through a huge organizational effort but are also the outcome of organization.

Today, big enterprises are the deployment centre of production processes. They concentrate their means of production, define and distribute their production tasks and constitute directly controllable teams of workers. At the same time, they adopt decentralized organization and management of their production resources. Production is presently organized as if the market power of an enterprise (and the coordination of functions and activities it succeeds in imposing on that market) constituted an economic power factor even more important than the power conferred on the firm by its own assets (scientific, technological, industrial and financial). However, one must not lose sight of the fact that the market power of that enterprise results from its financial capacity (ownership of financial assets and mobilization of capital) and from its informational potential, namely the whole set of scientific, technological, industrial, financial, commercial, political, sociological, etc. information to which an enterprise has access and/or distributes on the market. Once they have been associated, information and finance are used to build and manage groups of workers that are geographically dispersed and physically distant (investments devoted to inter-industrial cooperation relations; protection of technological assets; acquisition of scientific knowledge and design of novel commodities, coordination, etc.) (see Uzunidis and Boutillier, 1997; Laperche, 1998).

Moreover, technological innovations in the realm of information and communication enable that process to reach its ends and to be much more efficient (compared to the cost represented by the immobilization of a considerable volume of capital) than a giant production unit employing hundreds of workers. Discussions about 'networks' are focused on the flexibility (creation or dismantling of production units according to economic circumstances) resulting from the deconcentrated management of production. They also focus on the strengthening of the firm's capacity to acquire a considerable quantity of resources without having to invest in their formation. This flexible progressive coordination and innovation process compels the firm to invest in different types of technological and intellectual means for the acquisition and combination of endless flows of material and intangible resources. The concept of the 'knowledge economy' applied to enterprises produces the following result: the adaptation capacity and economic performance of the firm depend on its cognitive categories, codes of interpretation of the information itself; tacit expertise and procedures applicable to the solution of the problems it has to face (Dosi et al., 1999). Therefore, the task of the 'technostructure' is to identify the appropriate mix between the management of 'partnerships' and the development of its internal organizational tools.

Technological change is brought by, and generates in its turn, fundamental social and institutional changes. If we assume that economic cycles are accompanied by deep institutional changes, are we also entitled to consider that economic activity unrolls according to a regular movement comparable to natural processes (e.g. day and night or the seasons)? The obvious answer is no. Therefore, political and social relations are at the origin of economic change also in the realm of technological progress, labour organization, capital concentration and market development. In such conditions, long-lasting accumulation movements may be observed in the history of capitalism, but their periodicity is far from being mechanistic: capitalism evolves and changes under the pressure of political conflicts and the competition and struggles between capitalists. The resulting crises – characterized by a subsequent shrinking of wages and of investment opportunities – harbour many threats not only for enterprises but also for the social fabric of capitalism (see Schumpeter, 1979). At the same time, the situation thus created favours the emergence of new technologies, production methods and new social practices. Quite obviously, this presupposes the transformation of the social foundations necessary to create value and accumulation, otherwise the technological potential previously accumulated could never be fully used.

By accumulation foundations we mean the institutional forms of competition and cooperation between enterprises, capital owners and workers that enable, whenever necessary, social production relations to comply with the productive forces. Ownership relations and antagonisms between capital and labour as well as between enterprises and entrepreneurs must be reorganized in order to be in phase with the evolution of technologies, scientific and industrial knowledge, expertise and know-how. The organization of labour and of society as a whole must be altered in order to satisfy the requirements of the unremitting production of new merchant values and, in return, absorb and digest (always according to the merchant objectives of capitalism) scientific and technological progress. Economic crisis becomes unavoidable if the institutional changes brought to the regulation of economic activity fail to ensure a sufficient degree of compliance between the merchant relations relating to accumulation and private property and the scientific impact of production; hence the importance of global regulating policies.

2.2.2 Long-lasting movements and social change

When we refer to an economic cycle, we always assume a regular movement. This means that the behaviours of actors and the economic mechanisms are programmed to act in a stubborn manner according to the logic of 'growth–crisis–recovery–crisis ... '. Not only has this regularity never been observed in history, but the economic reasons for the crisis are diluted in highly contradictory and sometimes violent political relations (Freeman and Perez, 1986). On the one hand, the settlement of a crisis has never resulted from simple economic adjustments (as proposed since the 1980s, for example by international institutions such as the International Monetary Fund (IMF) or the World Trade Organization (WTO)) but rather from major social and political ruptures.

The title given by Schumpeter and his followers to each phase in the evolution of industrial capitalism (see Table 2.1) indicates that large-scale technological changes cannot be isolated from societal reorganization. The 'industrial revolution' is linked to the emergence of heroic entrepreneurs defined both as creators and owners of an enterprise and designers of new combinations of productive resources with a view to put on the market different types of technological and organizational novelties. Later, the extension of the production scale, markets and enterprises has placed 'affluent families' enjoying permanent revenues (assets, monetary savings and education) at the forefront of the economic stage. The 'bourgeois' cycle corresponds to the period of socialization of the capital of enterprises and the rise of 'big business' supported by the state and financial institutions (banks, stock exchange). The self-made man or heroic entrepreneur, starting from nothing, seizes the opportunities so frequently offered to him by the government in the form of investments or military/civilian orders. During the first half of the twentieth century, in the United States as well as in Europe, financial and industrial 'families' used to present themselves as those who had excelled in business. Puritanism and Darwinism emerged in the nick of time to justify the constitution of an elite with ever-growing assets (see also Bresser-Pereira, 2006, on this point).

Marx wrote in his *Capital* that credit becomes a vast social machinery whose aim is to centralize capital and demonstrated that, in particular since the middle of the nineteenth century, economic credit transactions have generated a banking system that operates independently from industry and trade. Thus a new market emerged, dealing with excess money capital that enabled bankers to exert control over production and trade enterprises. This provoked a widening of stock exchange transactions involving bonds, equities and other monetary securities. Stock exchanges became speculative and fraudulent centres; and later on, concentrated and centralized assets took on a financial and social character.

Speculation depersonalizes the owners of capital. Very often, scrip holders have never seen the good or the part of the assets these securities are supposed to represent and ignore the geographical location of the asset. Paul Lafargue wrote in 1909 that the whole contemporary economic development tends more and more to transform our society into a vast international gambling house in which bourgeois win and lose capital according to events they know nothing about, or that are totally unforeseeable, impossible to calculate and which they consider as being the result of chance and hazard. Had Schumpeter read Lafargue when he wrote in 1949 that by substituting a mere block of shares for the walls and machines of a factory, the joint-stock company, symbol of the success of capitalistic accumulation and of its future decomposition, socializes bourgeois mentality? Moreover, will it end by destroying the very roots of that system? (Schumpeter, 1979, pp. 179 and 212).

During the twentieth century, the 'mercantilist' and 'interventionist' cycles resulted from the opening of economies to global competition and to state interventions carried out to protect the property of the big industrialists, bankers and merchants. The increase observed in the scale of production is due to the overlap between the political and economic interests of the state and those of 'big business'. Starting from the 1930s, the domestic market of the major industrialized countries never ceased to increase. Improvements were made to the social division of labour; the social costs supported by enterprises and local and national communities as well as the savings made by the firms started to act in the same way as entrepreneurial opportunities of investments (namely, acquisition and combination of productive resources), like catalysers or barriers against accumulation. First, these factors had a pulling effect on growth; later, they turned into causes of depression because their potentialities had been overestimated.

The socialization of capitalistic production acquired such dimensions that the appropriation of the technological elements constituted by the firms is now more important than the mobilization of the capital necessary for their constitution. Through financial and commercial interactions, enterprises play the role of meeting places between knowledge and technologies they combine in order to fuel the innovation processes. In his day, Schumpeter was concerned by that situation as well as by the increasing size of capitalist enterprises. Monopolistic capitalism will provoke bureaucratization and the death of capitalism: merchant creativity will become the business of expert teams responsible for the planning of the manufacturing and marketing of new technologies and products. Without competition, there is no risk. In the absence of risk, there is no entrepreneurship. Without entrepreneurship, no innovation is possible. However, in his time Schumpeter could not imagine the future impact of finance on economic activities.

Finance succeeded in preserving the fundamental aspects of capitalism: it designed adequate tools so that capitalism was able to acquire the fruits of the socialization of production while dictating organization rules not only to enterprises but also to the economy as a whole. We suggest the term 'managerial' to describe the present phase of capitalism and offer two illustrations:

1. The reinforcement of the manager's economic and decisional power.
2. The securitization and marketization of any individual or collective asset. Economic policies geared towards worldwide privatization and financialization or through 'financial innovations' lead to a greater centralization of wealth (company buyouts or coalition of interests). This process also strengthens the role of asset managers.

By 'financial innovations' we mean the different methods that financial institutions discover every day in order to better increase and protect individual wealth. 'Financial derivatives' (futures swaps and other options), whose value is modelled on monetary or real assets, are used to cover the speculator should prices collapse. At the same time, speculators benefit from favourable developments. According to the World Bank, the global transactions including various financial products and performed by 'institutional investors' such as pension funds, insurance companies, investment funds and the business departments of banks, presently amount to more than $US2000 billion per day; that is 40 times the daily international volume of trade (World Bank, 2006).

The ascent of finance and managerial power may be illustrated by the theory of 'corporate governance' (Schleifer and Vishney, 1997). It describes the strong involvement in the daily operations of a big enterprise of the institutional (financial) investors that hold a considerable share of its capital. The short- and medium-term profitability of the capital invested by these holders is considered the most current criterion used to evaluate the technostructure's managers who hold the decision-making power in that enterprise. The exercise of that power requires numerous entrepreneurial competencies (Chandler, 1990). They are assessed in their turn by the managers' capacity to derive profits from the whole complex constituted by the 'momentum' of the enterprise and their success in incorporating external elements. The organization of an enterprise is modified by the following factors (Boutillier, 2005): increase in the social cost of innovation; strengthening of technological capabilities (in particular due to the introduction of information and biotechnologies); creation of a global market; invention of new production financing and managerial instruments; increase in the scale of capital mobilization through alliances and concentration. The enterprise delegates to a team of 'managers-entrepreneurs' the task of managing the centralized network responsible for the achievement of new productive combinations: big enterprises use their financial capacity to acquire and protect scientific knowledge and, quite often, to guide, according to their plans and economic objectives, the different choices and scientific research activities. In this perspective, the enterprise is supported by the innovation policy implemented in big countries. This innovation policy includes five major characteristics:

(a) reduction of public financing and resorting to contracts in the field of research where the 'return on investment' determines the selection of projects and the follow-up of the tasks performed;
(b) creation of institutions responsible for the commercialization of research inside universities: 'centres of excellence' are in charge of contract negotiations, patents application, licence agreements and the creation of technological enterprises;
(c) the elaboration of private statutes for those researchers who wish to participate in a research team or exit from it in order to create their own business, exploit their patents (or those of the centre employing them) or change their job;
(d) the development of institutions devoted to the financing of innovations and the creation of innovating enterprises, as for example venture capital;
(e) the support given to innovation networks connecting enterprises and universities.

'Knowledge and research' are acquiring a predominant place in economic and social development. Public managers and economists call upon academic institutions to participate in a 'third mission', namely to 'work with or like industry' (enterprises derived from universities, trilateral contracts – university, laboratories, enterprises – professional mobility between the public and private sector, e.g. Etzkowitz, 1999). It is true that universities not only concentrate a considerable share of basic research but are also rich in human resources already working in networks; they also serve as a basis for the formation of these resources to which the enterprise may have access at low cost.

The new phase of capitalism is not so much viewed according to its technological developments than by the new ways in which the production process is organized and put into operation. The industrial applications of science are the result of, but also the reason for, the accumulation, a way of reaching it and the cause of the emergence of crises. Present theories concerning networks, outsourcing, competition and innovation are based on an established principle: the advantages of the market, and a common conclusion: the market not only needs to be adapted, organized and regulated, but it also needs to be created and protected.

Today, production has taken on such dimensions that, from now on, the appropriation of technological elements concentrated in big enterprises is cheaper than the mobilization of capital necessary for their formation. Big enterprises, as major nodes of economic activities, are evolving into poles which are used for the convergence of the scientific and technological knowledge they combine in order to fuel their innovation process. To go from the previous stage of production concentration to the present phase of contractual integration of a centralized property, capitalism invented a new accumulation framework. Economic policies based on 'monopoly contestability', liberalization, flexible labour management and international integration have to a certain extent succeeded in destroying old capital. These policies have created the conditions of organization and economic control favourable to company managers and public technocrats (Uzunidis, 2004). This situation can be observed at two levels: (1) the financial regulating power that defines the selection criteria of the scientific and innovation programmes; (2) industrial and innovation policy. Initially, public intervention relates to the offer of scientific resources starting from their transfer from the public to the private sector. Present innovation policies enable enterprises to have access to all the scientific means of research, development and application and to make technological choices for the elaboration of new production mechanisms and processes, new industrial goods and services. The second facet of such a policy is the formulation of a framework favourable to the diffusion–appropriation of innovation through a review of competition rules (questioning of monopoly situations on the markets of goods and services, finance and labour), the introduction of market principles and acquisition in all the domains of scientific activity; the formulation of well-targeted innovation (civilian and military) programmes. Here, 'private' managers are associated with 'public' managers to give full meaning to the 'network economy' (Laperche et al., 2006). The question is to know whether the contemporary 'network economy' is a vector of a sustainable growth. On the one hand, the size of the markets (with their costly and quick renewal) and the resulting profit opportunities look ridiculous compared to their volumes of capital accumulated; on the other side, the increasing degree of wealth concentration limits the exploitation of the scientific, technological and economic potentialities, in more general terms. No wide-scale accumulation movement, no cycle will be triggered as long as demand remains weak and finance punctures an increasing value surplus to turn it into a rent.

However, we wonder what type of demand will succeed in amortizing technological capital. Individual or collective needs? Merchant or non-merchant? The economy, as a reality and as a science, is subjected to the imperatives of policies and theories based on the individual, divisible demand, scarcity, marginal utility and marginal productivity. However, the economic and social potential stemming from scientific and technological progress is of a collective nature, thus opposed to individual criteria of profitability: health, nutrition and biotechnology, education and information and communication technology, environment and renewable energy. Their profitability may only be evaluated by using collective consumption criteria. We could imagine the formation of a common investment fund (public expenses and enterprises' contributions) dedicated to the production of these collective goods and services. These investments would only be profitable when they satisfy a collective demand for education, health and the environment. As a consequence, profitability would have no relation to the efficiency of each invested unit of capital. The profitability criteria would concern this collective, global and compulsory investment. The profitability criteria would be defined *ex post*: free access to all forms of knowledge, to health services, to the necessities of life (air, water). In face of the current barriers to accumulation, we may consider the fact that economic evolution in future years will depend on the width of global non-merchant sectors.

Institutional innovation, change, transition, crisis: such is the modus operandi of economic reality. Obviously such technological and social innovations help us to understand the dynamic of long periods in dominating economies.[1] The question is to know how economies may recover over time once their mode of development has been exhausted. The answers given by economists to that fundamental question remain insufficient.

Note

1 Freeman and Louçã made a well-documented study of the evolution of capitalism and stressed in their conclusion that it cannot be limited to a mere economic process (Freeman and Louçã, 2001).

References

Boutillier, S., *Travail et entreprise. Règles libérales et global management* (Paris: L'Harmattan, 2005).
Bresser Pereira, L. C., 'Professional's Capitalism and Democracy', in B. Laperche, J. Galbraith and D. Uzunidis (eds), *Innovation, Evolution and Economic Change* (Cheltenham: Edward Elgar, 2006), pp. 17–37.
Chandler, A., *Scale and Scope. The Dynamics of Industrial Capitalism* (Cambridge: Harvard University Press, 1990).
Dosi, G., R. Nelson and S. Winter (eds), *The Nature and Dynamics of Organizational Capabilities* (Oxford: Oxford University Press, 1999).
Etzkowitz, H., *The Second Academic Revolution, MIT and the Rise of Entrepreneurial Science* (London: Gordon & Breach, 1999).
Freeman, C. and F. Louçã, *As Time Goes By. From the Industrial Revolutions to the Information Revolution* (Oxford: Oxford University Press, 2001).
Freeman, C. and C. Perez, *The Diffusion of Technical Innovations and Changes of Techno-Economic Paradigm* (Sussex: SPRU, 1986).
Galbraith, J. K., *The New Industrial State* (Boston, Mass.: Houghton Mifflin, 1967).
Habermas, J., *La technique et la science comme idéologie* (Paris: Gallimard, 1973).
Karpik, L., 'Le capitalisme technologique', *Sociologie du travail*, 1 (1972), 2–34.
Kondratiev, N. (1935), *The Long Wave Cycle* (New York: Richardson & Snyder, 1984).
Lafargue, P. (1909), *Le déterminisme économique de Karl Marx* (Paris: L'Harmattan, 1997).
Laperche, B., *La firme et l'information* (Paris: L'Harmattan, 1998).
Laperche, B., *L'entreprise innovante et le marché* (Paris: L'Harmattan, 2005).
Laperche, B., J. K. Galbraith and D. Uzunidis (eds), *Innovation, Evolution and Economic Change. New Ideas in the Tradition of Galbraith* (Cheltenham: Edward Elgar, 2006).
Marx, K. (1867), *Le Capital* (Paris: Editions sociales, 1976).
Marx, K. (1857), *Grundrisse* (Paris: Anthropos, 1977).
Nef, J., 'Essence de la civilisation industrielle', in *Hommage à L. Febvre* (Paris: Armand Colin, 1953), pp. 93–114.
Robinson, J., 'What are the Questions?', *Journal of Economic Literature*, 15 (1977), 1318–39.
Schleifer, A. and W. Vishney, 'A Survey of Corporate Governance', *Journal of Finance*, 2 (1997), 471–517.
Schumpeter, J., *Business Cycles* (New York: McGraw-Hill, 1939).
Schumpeter, J., *Théorie de l'évolution économique* (Paris: Dalloz, 1935).
Schumpeter, J. (1942), *Capitalisme, socialisme et démocratie* (Paris: Payot, 1979).
Uzunidis, D. and S. Boutillier, *Le travail bradé* (Paris: L'Harmattan, 1997).
Uzunidis, D., *L'innovation et l'économie contemporaine. Espaces cognitifs et territoriaux* (Brussels: De Boeck, 2004).
World Bank, *Global Development Finance 2006* (Washington: World Bank, 2006).

3
Macroeconomic Policy, Investment and Innovation

Malcolm Sawyer

3.1 Introduction

A, perhaps the, major difference between schools of thought in macroeconomic analysis arises from the relationship between the demand side and the supply side. Say's Law is, of course, often summarized as 'supply creates its own demand' and the view that the level of economic activity is determined by the supply potential, and that demand has to adjust to underpin that level of supply. There have been long debates over whether this is a feature of the short run as well as the long run. The separation between demand and supply, and the view that there is a predetermined growth of supply, is reflected in the following quote from the now Governor of the Bank of England: 'if one believes that, in the long-run, there is no trade-off between inflation and output then there is no point in using monetary policy to target output. ... [You only have to adhere to] the view that printing money cannot raise long-run productivity growth, in order to believe that inflation rather than output is the only sensible objective of monetary policy in the long-run' (King, 1997, p. 6). Similar sentiments are often expressed with respect to fiscal policy, and the more general view that the level of demand has no long-run effects on the levels of output and employment in an economy. The vertical aggregate supply curve is a regular feature of macroeconomic analysis, and is an embodiment of the view that the level of economic activity (in the long term) is supply-determined (with the implicit assumption being made that the vertical aggregate supply curve is not influenced by the time path of aggregate demand.

This approach stands in contrast to the broadly Keynesian notion that the level of economic activity is set in the short and the long run by the level of aggregate demand. But further that the level of demand has effects on the potential supply, and supply adapts to demand. This is perhaps most evident for capital where investment responds to demand (actual and prospective), and that generates changes in the capital stock, and hence the future supply potential. In the neoclassical analysis, the level of investment is taken to passively adjust to savings, and in turn savings reflect households' desire to distribute their spending power over their lifetime. Savings decisions by households in effect drive the rate of capital formation. But the effective size of the labour force can through a variety of routes also respond to the level of demand, for example changes in participation rates, migration. It is not only that future supply potential depends on demand, but that also implies that the supply side of the economy is path-dependent on the path of aggregate demand.

Another, and particularly important, aspect of the evolution of supply potential comes from the invention and innovation of products and production processes and the growth of productivity. The literature of invention and innovation has long recognized the impact of economic activity on invention and innovation, and vice versa. 'As scholars from Marshall...to Kuznets...have recognized, economic activity changes knowledge directly and indirectly and every change in knowledge opens up the conditions for *changes in activity* and *thus further changes in knowledge, ad infinitum*, and in quite unpredictable ways' (Metcalfe, 2001, p. 570). The notion of 'learning by doing' and the productivity growth which comes from higher levels of economic activity are clearly another route by which the level of demand through its influence on the level of economic activity has further effects on the supply side of the economy.

This view suggests two points. First, it links with the view that the future is inherently uncertain (in the Knightian and Keynesian senses). In contrast, a rational expectations perspective is built on a combination of massive computational power by individuals and an essentially knowable future (with risk but not uncertainty recognized). A great deal of macroeconomic analysis has been built on rational expectations, and the 'new consensus in macroeconomics view' which we briefly outline in the next section is only the most recent evidence of this. Second, it reinforces the path-dependency view of the economy. This is more than merely that a current event may have long-lasting effects which eventually die away, the view typically expressed in the hysteresis literature with regard to unemployment. It is that the future is built on the past and present and that decisions taken in the present will impact on the future course of the economy through 'lock-in' effects, effects on investment and capital accumulation and the processes of technical change.

In this chapter we consider how two approaches to macroeconomic analysis incorporate (or fail to incorporate) the links between the demand side and the supply side and the treatment given to technical change. From that discussion we draw conclusions on the ways in which the two approaches provide a perspective on macroeconomic policy.

3.2 The new consensus to macroeconomics approach

The currently dominant approach to macroeconomic thinking, particularly with regard to policy making and the roles of monetary and fiscal policies, is generally referred to as the 'new consensus in macroeconomics' (NCM). In this section we briefly outline the NCM with an emphasis on the way it treats the relationship between supply and demand, and also how the NCM (implicitly) treats the processes of invention and innovation. Further, we wish to ask whether the NCM is a 'safe' basis on which to base macroeconomic policy when the economy is characterized by new products and processes and technological change.

The key features of the NCM start from the major role played by the supply-side-determined equilibrium level of unemployment (the 'natural rate' or the non-accelerating inflation rate of unemployment, the NAIRU). It generally neglects the role of aggregate or effective demand and fiscal policy other than in the short run. Monetary policy is elevated at the expense of fiscal policy and is targeted towards the control of inflation. The NCM can be described succinctly in the following three equations (see, for example, McCallum, 2001; Arestis and Sawyer, 2004a):[1]

$$Y_t^g = a_0 + a_1 Y_{t-1}^g + a_2 E_t(Y_{t+1}^g) - a_3[R_t - E_t(p_{t+1})] + s_1 \quad (3.1)$$

$$p_t = b_1 Y_t^g + b_2 p_{t-1} + b_3 E_t(p_{t+1}) + s_2 \quad (3.2)$$

$$R_t = (1 - c_3)[RR^* + E_t(p_{t+1}) + c_1 Y_{t-1}^g + c_2(p_{t-1} - p^T)] + c_3 R_{t-1} + s_3 \quad (3.3)$$

with $b_2 + b_3 = 1$, where Y^g is the output gap, R is nominal rate of interest, p is rate of inflation, p^T is inflation rate target, RR^* is the 'equilibrium' real rate of interest, that is the rate of interest consistent with zero output gap which implies from Equation (3.2) a constant rate of inflation, s_i (with $i=1$, 2, 3) represents stochastic shocks, and E_t refers to expectations held at time t. Equation (3.1) is the aggregate demand equation with the current output gap determined by past and expected future output gap and the real rate of interest. Equation (3.2) is a Phillips curve with inflation based on current output gap and past and future inflation. Equation (3.3) is a monetary-policy rule in which the nominal interest rate is based on expected inflation, output gap, deviation of inflation from target (or 'inflation gap'), and the 'equilibrium' real rate of interest. The lagged interest rate represents interest rate 'smoothing' undertaken by the monetary authorities, which is thought to improve performance by introducing 'history dependence' (see, for example, Rotemberg and Woodford, 1997; Woodford, 1999). Variations on this theme are used. For example, interest rate 'smoothing' in Equation (3.3) is often ignored, as is the lagged output gap variable in Equation (3.1) so that the focus is on the influence of expected future output gap in this equation. There are three equations and three unknowns of output, interest rate and inflation, giving a determinate solution.

The use of output gap, that is the difference between the actual level of output and some notion of trend or normal output, rather than the level of output per se in this formulation is significant. The trend output can be estimated in many ways, and it is implicit in the estimation of trend that the 'average' level of output reflects supply-side conditions, and that this 'average' level can be taken as reflecting some form of supply-side equilibrium. Further, it is implicit that the level of demand does not influence the 'average' level of output. While it is recognized that the level of demand influences actual output, this effect of demand is assumed to 'average' out at zero. It is clearly the case that the trend can only be estimated after the event, and contemporaneous estimates of output gap have to rely on extrapolation of past trends.

The assumption being made is equivalent to the assumption of the constancy of the 'natural rate of unemployment' in that it is assumed that it is always the case that a zero output gap would be consistent with a constant rate of inflation. The assumption of the lack of impact of demand on the future course of the economy is particularly important. Monetary policy is perceived to operate through its impact on demand (which may come through a variety of channels) which feeds through to the level of economic activity and then to the rate of inflation. The link from demand to inflation may be problematic. Here we emphasize that monetary policy operates through interest-sensitive expenditures, such as investment. Insofar as interest rates influence investment, then they influence the future capacity (and trend output) of the economy.

Let us look briefly at the derivation of the first two equations. Equation (3.1) is typically derived from optimization of expected lifetime utility subject to a budget constraint (see, for example, Blanchard and Fischer,

1989, Ch. 2). Households and firms are assumed to have perfect foresight, and know the current and future values of wages and rental rates. A condition, often referred to as a no-Ponzi game (NPG) condition, is imposed which 'prevents families from choosing such a path [with higher and higher levels of borrowing], with an exploding debt relative to the size of the family. At the same time we do not want to impose a condition that rules out temporary indebtedness. A natural condition is to require that family debt not increase asymptotically faster than the interest rate' (Blanchard and Fischer, 1989, p. 49).

Three features of this approach should be noted. First, the NPG condition leads to the implication that lifetime consumption is equal to lifetime income (each suitably discounted). At the individual level, this comes from a combination of a non-satiation assumption along with a no final debt condition. Second, the income of the individual depends on labour supply at the given real wage, and the implicit assumption that the individual is able to actually supply their labour. There is a full employment assumption. The combination of these two features means that at the aggregate level there is the equivalent of Say's Law: potential supply (of labour) leads to actual supply of labour, and the resulting income is fully spent.

Third, the consumption decision is made at the level of the household or family under perfect foresight. Objections made be raised to the notion of intertemporal optimization along the lines of 'unrealism' in terms of the information on the future levels of income, interest rates, etc. which are required and the computational requirements to solve the optimization problem. No consideration is given to genuine uncertainty about the future, to learning and the change in household membership. A certain (or certainty equivalent) future is postulated. Significantly there is little room for learning in this process. A further complication arises from whether the optimization is carried out at the individual level or the household level. If the decision is at the individual level, then some consideration should be given to income sharing within a household. If the decision is presented as being made at the household level (e.g. as in Heijdra and van der Ploeg, 2002) then there should be some recognition of changing household composition – children grow up, households split, etc.

If the future were essentially like the past, and future events could be thought of in terms of risk (that is the probability distribution of future events is known), then there would be some appeal in envisaging that individuals base their current decisions on their expectations of future events. A model in which individuals optimize their lifetime pattern of consumption with firmly based expectations on future income requires individuals to have the computational facilities to make the decisions and for the future to be predictable. Although the future is inherently unknowable, the one fact about the future which we do know is that it will be different from the past. When economies and societies are evolving and when innovation and technical change are occurring, the past and present do not provide sufficient guidance for the future to enable optimizing decisions to be made. The time path of the economy will be influenced then by the pace of innovation and technical change, but in ways which individuals cannot foresee.

There are numerous features of this NCM approach which serve to reinforce the separation of the evolution of the supply side from the demand side, and which also serve to limit the roles of monetary and fiscal policy. Here we highlight three.

The first is the perfect capital market assumption that individuals and firms can borrow or lend as much as they wish at the prevailing rate of interest (subject to conforming to an overall lifetime budget constraint). This implies the absence of credit rationing under which some individuals were credit constrained in their expenditure decisions. This would mean that the only effect of monetary policy would be a 'price effect' as the rate of interest is changed. The parts of the transmission mechanism of monetary policy, which involve credit rationing and changes in the non-price terms on which credit is supplied, would be excluded by assumption. More significantly the absence of credit rationing means that individuals who currently have fallen on hard times are able to maintain their consumption levels by borrowing against future income.

Second, there is no mention of banks in this analysis. It has been noted that in the major text of Woodford (2003) banks make no appearance in the index. Since banks and their decisions play a considerable role in the transmission mechanism of monetary policy, and further that decisions by banks as to whether or not to grant credit play a major role in the expansion of the economy (in the sense that a failure of banks to supply credit would imply that expansion of expenditure cannot occur), there is a disjuncture between this analysis and the role of monetary policy.

Third, the role for investment is to match savings, and there is no independent investment function. The basic analysis (cf. Woodford, 2003, Ch. 4) is undertaken for households optimizing their utility function in terms of the time path of consumption. 'One of the more obvious omissions in the basic neo-Wicksellian model developed in Chapter 4 is the absence of any effect of variations in private spending upon the economy's productive capacity and hence upon supply costs in subsequent periods' (Woodford, 2003, p. 352). Investment can then be introduced in terms of the expansion of the capital stock, which is required to underpin the growth of income. In effect the future path of the economy is mapped out, and consequently the time path of the capital stock. Investment ensures the adjustment of the capital stock to that predetermined time path. There is then by assumption no impact of the path of the economy on the capital stock. Further, there can be no allowance for the interrelationships between investment and technical change and innovation. The ways in which investment (notably in research and development) can influence the pace of technical change and the ways in which the prospects for technical change create investment opportunities are both ignored.

In this approach, there is a corresponding government budget constraint. This takes the form of 'the government's *intertemporal budget constraint*. ... It states that *the current level of debt must be equal to the present discounted value of primary surpluses*. If the government is currently a net debtor, it must intend to run primary surpluses at some time in the future' (emphasis in original, Blanchard and Fischer, 1989, p. 127).

3.3 A Kaleckian analysis

It is commonplace to observe that the level of economic activity is demand determined in the short run, and that fluctuations in the level of economic activity arise from fluctuations in demand. The Kaleckian analysis views significance of the role of aggregate demand as more extensive than that. Specifically, the absence of market-based forces leading the level of demand into line with available supply is one basic tenet of a Kaleckian analysis and hence inadequate aggregate demand can be a long-term phenomenon. Further, the evolution of the supply potential of the economy in terms of the available workforce, the size of the capital and the growth of factor productivity are all strongly influenced by the time path of the level of demand. This is most evident for the growth of the capital stock, where investment expenditure is strongly influenced by the level of economic activity, but it would also be relevant for the evolution of the effective labour force.

The level of aggregate demand depends on the sum of intended consumer demand, investment demand and government expenditure plus the net trade balance. Since the propensity to consume depends on income source (wages versus profits) and investment is influenced by profitability for a variety of reasons, the distribution of income between wages and profits plays a significant role in the determination of aggregate demand. The level of economic activity itself (as reflected in capacity utilization, level of unemployment) impacts on the level of aggregate demand. Investment decisions involve commitments and rewards which extend far into the future, and when the future is viewed as inherently uncertain investment decisions cannot be approached through optimization under full information about a predetermined future. Investment decisions (along with many others) are not then approached through seeking to set up some optimization problem from which first-order conditions are derived to be used for an investment equation. Recent and current experience along with views about the future will have a strong influence on investment. Hence investment is path-dependent, and specifically is influenced by the path taken by demand and economic activity, and reflected in variables such as profitability and capacity utilization. There is no sense in which the future time path of the capital stock can be seen as predetermined by relative prices (as in the neoclassical approach). When investment and hence the evolution of the capital stock are path-dependent, then macroeconomic policies have an influence on investment, and thereby on the evolution of the supply side of the economy as investment adds to the capital stock.

The variables which impact on investment are particularly significant for the effects or otherwise of monetary policy. In the Kaleckian tradition we postulate that profits (current and anticipated) and capacity utilization (level and change) are significant (see, for example, Lavoie, 1992, Ch. 6; Sawyer, 1982, pp. 99–104, 155–6; Mott, 2003). The inclusion of profits comes from a range of considerations. Current profits provide a potential pool of funds for the internal financing of investment. They may also be used as a signal of future profitability. Capacity utilization clearly relates to the idea that firms undertaking investment in order to add to the capital stock are able to produce higher levels of output in the future. Underutilized capacity would dampen the need to undertake investment for that purpose.

Investment expenditure has to be financed, and hence the condition under which finance is made available is significant. Kalecki provided the essence of the argument here.

> [T]he possibility of stimulating the business upswing is based on the assumption that the banking system, especially the central bank, will be able to expand credits without such a considerable increase in the rate of interest. If the banking system reacted so inflexibly to

every increase in the demand for credit, then no boom would be possible on account of a new invention, nor any automatic upswing in the business cycle.... Investments would cease to be the channel through which additional purchasing power, unquestionably the *primus movens* of the business upswing, flows into the economy. (Kalecki, 1990, p. 489)

The cost of finance may also have some influence on investment decisions, though the relationship between the perceived rate of profit and the long-term rate of interest would be particularly relevant.

In Sawyer (2002) a model was developed which had some features particularly relevant for this discussion (see also Arestis and Sawyer, 2005). An inflation barrier was derived from the interaction of wage- and price-setting behaviour. This barrier had some resemblance to a NAIRU (non-accelerating inflation rate of unemployment) in the sense of being a supply-side equilibrium. But it differs from the conventional NAIRU in two important respects. First, the inflation barrier, unlike the NAIRU, is not seen as a labour market phenomenon, and cannot be shifted through labour market 'reforms'. The inflation barrier is viewed as the levels of unemployment and capacity utilization which hold inflation in check and is seen in terms of struggles over the distribution of income. Second, the position of the inflation barrier crucially depends on the size of the capital stock, and hence the inflation barrier is perpetually shifting as the capital stock changes through investment.

From that model we derived three implications. First, there is an inevitable path dependence of the inflation barrier, with the path dependence coming from the path of investment. Second, the path of demand influences the path of investment (through capacity, profitability and prospects for the future), and thereby the apparent supply-side equilibrium. The path of demand influences the supply side of the economy. Third, although there is a supply-side equilibrium (expressed as the inflation barrier) there are no systematic forces which lead the level of economic activity to move towards that equilibrium position. The level of economic activity is determined by the level of aggregate demand, and there are no market forces which would ensure that the level of aggregate demand adjusts to the supply-side equilibrium. There may be government policies, such as monetary policy, which seek to do so, but not market forces.

Investment is related to the level of aggregate demand, with the latter depending on capacity utilization and profitability (Kalecki, 1943). Hence, the evolution of the capital stock depends on the time path of the level of aggregate demand. Higher levels of aggregate demand lead to more investment and over time a larger capital stock. Consequently, the significance of investment in our approach arises through the impact of investment on the capital stock, and thereby on the inflation barrier. But further investment is seen as influenced by a range of factors including capacity utilization and profitability, and the level of aggregate demand will impact on those factors. Hence the capital stock is not only path-dependent but also depends on the path of aggregate demand.

3.4 Perspectives on macroeconomic policies

The widespread approach to macroeconomic policy in industrialized countries, exemplified by the Stability and Growth Pact of the Economic and Monetary Union, combines a focus on monetary policy aimed to target inflation and a passive fiscal policy. The NCM outlined above is used to support that general policy stance. Monetary policy, often conducted by an 'independent' central bank, takes the form of interest rate policy, through which the general level of demand is influenced, which in turn feeds impacts on inflation. This can be justified on the basis that monetary policy (and more general demand management policies) have no lasting effects on other economic variables such as (un)employment, economic growth and the exchange rate. The expectations-augmented Phillips curve has the attribute that there is a level of economic activity (such as the 'natural rate of unemployment') consistent with a constant rate of inflation. Hence if inflation is to be constant (and specifically if the target rate of inflation is to be achieved), then the level of economic activity has to be at this 'supply-side equilibrium'. It is further assumed that this 'supply-side equilibrium' is unaffected by the time path of the level of economic activity (and of demand). Thus path dependence (of the supply-side equilibrium) is ruled out by assumption.[2] It is assumed that monetary policy (in the form of the policy interest rate) can achieve the desired level of demand (in the medium term that is one compatible with the 'supply-side' equilibrium). It is generally postulated that there is a 'natural' rate of interest (akin to that postulated by Wicksell) which equates aggregate demand with aggregate supply (or equivalently which equates *ex ante* savings and investment at the 'supply-side' equilibrium level of economic activity).

The general Kaleckian view is that the level of aggregate demand will be little affected by the level of interest rates. Such doubts stem from the view that long-term interest rates were seen as relevant for many decisions but there were rather small movements in such long-term rates. 'The relative stability of the long-term rate of interest is generally known.... It seems unlikely that changes in the long-term rate of interest of the order of those noticed ... can influence investment activity' (Kalecki, 1990, pp. 296-7). It would also be that a substantial fall in the rate of interest would be necessary to have a significant effect on investment (cf. Kalecki, 1990, p. 403). Aggregate demand may then be rather unresponsive to interest rates, which would imply that the changes in interest rates required to generate significant changes in output and inflation may be rather large (that is of the order of several percentage points). Further, the effects of changes are likely to be a rather unpredictable way of generating changes in aggregate demand. As we pointed out in Arestis and Sawyer (2004a), empirical work (often undertaken within central banks) on the impact of interest rates on demand and inflation suggested that insofar as there was an effect on demand this came through effect on investment rather than consumer demand.

The general argument that monetary policy is rather ineffectual was widely accepted during the 1950s and 1960s. The position has now been generally reversed and monetary policy is widely used. Swimming against that mainstream tide, we argue that monetary policy (in the form of interest rates) is generally ineffectual. But the main argument we want to put here is that insofar as interest rate variations have effects on the level of demand it will be through interest-sensitive expenditures such as investment. Thereby monetary policy has long-lasting effects (though those effects may be rather small). Once it is recognized that interest rates can have some effects on the supply capacity of the economy, the case for an 'independent' central bank operating a monetary policy focused solely on inflation is very much undermined. Monetary policy would then be seen to have effects on the short-run level of demand and also on the long-run supply potential of the economy.

When consideration is given to technical change and innovation, the general picture is rather similar. The effects of monetary policy on demand may be rather small, but insofar as there are effects, there can be effects on technical change and innovation. The effects of investment and the general level of economic activity on technical change would mean that there are long-lasting effects of monetary policy.

The point was made above that the NCM incorporates the view from the neoclassical growth model that savings arise from household decision making and that savings determine capital accumulation. The contrast can be made with the Kaleckian approach in which there is an independent investment function, and additions to the capital stock are path-dependent. There is no notion of a pre-existing optimal capital stock depending on relative prices towards which adjustment is made through investment. It is rather that the 'long run is a collection of short runs', since investment depends on the level of and change in economic activity. Hence when economic activity is high, investment is relatively high and more additions are being made to the capital stock. Since the level of economic activity depends on the level of aggregate demand, the additions to the capital stock depend on the level of demand. The size of the capital stock is obviously an important ingredient in the supply capability of the economy. In the medium term at least the additions to the capital stock which take place cannot be ignored in terms of the supply side of the economy. Insofar as interest rates do have an impact it could come through investment. But if higher interest rates are used to reduce inflation and work through an impact on investment, then the future capital stock is lower than it would have been, and worsens the prospects for inflation insofar as the inflation barrier shifts. The subsequent inflation barrier becomes more of an obstacle, and the inflationary pressures corresponding to a particular level of demand greater as the supply potential would be lower.

In a Kaleckian analysis, fiscal policy retains a key role in supporting the level of aggregate demand. We have discussed elsewhere (Arestis and Sawyer, 2004b, c) the various objections raised against fiscal policy and found them wanting. The essence of our argument there was that fiscal policy has been dismissed by many mainstream economists by using an analytical framework in which there was no deficiency of private aggregate demand, and hence no need for fiscal policy. In contrast, the Kaleckian analysis is firmly based on the view that private aggregate demand is often insufficient to support full employment. The current vogue is to assert the role of interest rates in securing a level of aggregate demand in line with available supply, and notably the role of the 'natural rate of interest'. The Kaleckian analysis largely dismisses those arguments. Hence there is a remaining need for fiscal policy to influence the level of aggregate demand.

The particular argument put here is that the case against the use of fiscal policy (and notably budget deficits) is based on the view that there is not a general deficiency of aggregate demand. In turn that is based on the idea that individuals will spend all their (potential) income: that is, Say's Law operates. Individuals are optimizing and knowing their future income will spend up to the limit of their income. Further, as indicated above, the savings which individuals make in any specific period will flow into investment (by assumption in the NCM model). Technical change and innovation not only mean that the future will surely be different from the past, they also confirm that essential unpredictability and unknowability of the future. Thus grave doubts have then to be cast on the use of an analysis which is firmly based on the predictability of the future.

3.5 Concluding remarks

The prevailing approach to macroeconomic policy is based on a number of propositions, of which we highlight two related ones. First, there is a classical dichotomy between the real supply side of the economy and the demand side of the economy. In the short run the level of economic activity (relative to the supply-side equilibrium level) is determined by the level of demand, but demand moves towards the supply-side equilibrium (guided by the central bank through their interest rate setting). The supply-side equilibrium is left untouched by what happens to the level of demand and the level of economic activity. Second, the future path of the economy is also set by the supply side of the economy. Individuals are well informed on that future path, and make decisions on consumption and savings based on their expectations of future income. The savings decisions flow into investment, and capital accumulation depends on the savings decisions of households. There is then no inherent deficiency of demand.

We have argued here that when the processes of innovation and change are taken seriously then this approach to macroeconomic policy is severely undermined, essentially on two counts. First, the classical dichotomy breaks down. Decisions made by firms with regard to investment in capital equipment and in research and development impact on the supply potential of an economy. Further, demand influences investment, and also influences the rate of productivity growth, for example through learning by doing. Second, individuals cannot optimize with full information on the future when the future is inherently unknowable. The conclusion reached from the use of optimizing models with full information that there would be no inherent inadequacy of aggregate demand would not be sustained in an uncertain world. Fiscal and other macroeconomic policies are then required to address these potential problems of inadequate aggregate demand.

Monetary policy in most countries is undertaken by the central bank and targets the rate of inflation, and this is based on the assumption that the underlying level and rate of growth of output will be unaffected by demand. Fiscal policy has been much downgraded on the basis that there is no persistent deficiency of aggregate demand problem, and any such issue can be dealt with by the setting of an appropriate interest rate (so that the level of demand will be consistent with the supply-side equilibrium). We have argued here that output and employment in the long as well as the short run are set by the level of demand, and that the supply potential of the economy (including productivity) is influenced by the level of demand. Further we have argued that there is often a deficiency of demand issue. The policy implications of our discussion are that monetary policy has to pay regard to the output, demand and supply implications of interest rates, and not use interest rates solely to target inflation, and that fiscal policy is required to address issues of deficient aggregate demand.

Notes

1 See Arestis (2007) for an extended six-equation open-economy model based on the NCM: the model in the text reflects the key elements of the NCM relevant for the present discussion.
2 An alternative view would be that demand fluctuates around the supply-side equilibrium, sometimes above, sometimes below, and any effects of demand on the supply-side equilibrium balance out at zero.

References

Arestis, P., 'What is the New Consensus in Macroeconomics?', in P. Arestis (ed.), *Is There a New Consensus in Macroeconomics?* (Basingstoke: Palgrave Macmillan, 2007).
Arestis, P. and M. Sawyer, 'Can Monetary Policy Affect the Real Economy?', *European Review of Economics and Finance*, 3(3) (2004a), 9–32.
Arestis, P. and M. Sawyer, 'Reinventing Fiscal Policy', *Journal of Post Keynesian Economics*, 26(1) (2004b), 4–25.
Arestis, P. and M. Sawyer, 'On Fiscal Policy and Budget Deficits', *Intervention, Journal of Economics*, 1(2) (2004c), 61–74.
Arestis, P. and M. Sawyer, 'Aggregate Demand, Conflict and Capacity in the Inflationary Process', *Cambridge Journal of Economics*, 29(6) (2005), 959–74.
Blanchard, O. J. and S. Fischer, *Lectures on Macroeconomics* (Cambridge, Mass. and London: The MIT Press, 1989).
Heijdra, B. J. and F. van der Ploeg, *Foundations of Modern Macroeconomics* (Oxford: Oxford University Press, 2002).
Kalecki, M., 'The Determinants of Investment', in *Studies in Economic Dynamics* (London: Allen and Unwin, 1943), pp. 171–205.
Kalecki, M., *Capitalism: Business Cycles and Full Employment, Collected Works of Michal Kalecki*, vol. I (ed. by J. Osiatynski) (Oxford: Clarendon Press, 1990).
King, M., Lecture given at London School of Economics (1997).
Lavoie, M., *Foundations of Post-Keynesian Economic Analysis* (Aldershot: Edward Elgar, 1992).

McCallum, B. T., 'Monetary Policy Analysis in Models without Money', *Federal Reserve Bank of St. Louis Review*, 83(4) (2001), 145–60.
Metcalfe, J. S., 'Institutions and Progress', *Industrial and Corporate Change*, 10(3) (2001), 561–85.
Mott, T., 'Investment' in J. E. King (ed.), *The Elgar Companion to Post Keynesian Economics* (Cheltenham: Edward Elgar, 2003), pp. 205–10.
Rotemberg, J. J. and M. Woodford, 'An Optimization-based Econometric Framework for the Evaluation of Monetary Policy', *NBER Macroeconomics Annual 1997* (Cambridge, Mass.: National Bureau of Economic Research, 1997), pp. 297–346.
Sawyer, M., *Macroeconomics in Question* (Brighton: Harvester Wheatsheaf, 1982).
Sawyer, M., 'The NAIRU, Aggregate Demand and Investment', *Metroeconomica*, 53(1) (2002), 66–94.
Woodford, M., 'Optimal Monetary Policy Inertia', NBER Working Paper Series, 7261 (Cambridge, Mass.: National Bureau of Economic Research, 1999).
Woodford, M., *Interest and Prices: Foundations of a Theory of Monetary Policy* (Princeton: Princeton University Press, 2003).

4
Doctrinal Roots of Short-Termism
James E. Sawyer

4.1 Introduction

This chapter examines short-termism in the United States and connects it with doctrinal antecedents in the received neoclassical theory. Specifically, it argues that capital in wealthy, post-industrial societies must be defined relative to preferred outcomes for which capital assets are to be deployed. This is consistent with the neoclassical profit 'lacuna' that in disequilibrium conflates the 'productivity' of asset placements by capitalists with the 'non-productivity' of asset placements by rentiers. Also conflated are the rewards to the respective classes of capitalists and rentiers. Consequently, the unbridled pursuit of self-interest in wealthy, financially sophisticated economies leads frequently to rent-seeking, 'pseudo-capitalist' behaviours in which institutions and individuals 'untether' themselves from the conventional capitalist relationship: holding capital 'patiently' over the longer term. In sum, the pursuit of self-interest may not articulate with the attainment of the common good when financial innovations enable pseudo-capitalists to shed asset positions prematurely.

Short-termism can be addressed thoroughly only by addressing policy implications of the profit lacuna, thereby setting the rate of profit on capital exogenously, through the political process. Distinction must be made between capital assets deployed productively and other assets hoarded into non-productive use through rent-seeking strategies. In wealthy post-industrial societies, this productive/non-productive distinction lies increasingly within the realm of choice, rather than necessity. Therefore, before the profit premium can be set, society must specify its favoured pathway into the future, and then specify the menu of productive capital assets essential to fulfil its plan. These assets then qualify for preferential tax and regulatory treatment.

Short-termism cannot be remedied by conventional government oversight or business sector resolve alone, without clarifying the nature of what capital investment shall be and the differential reward that shall flow to those holding it. Often short-termism in the United States takes the form of financial manipulation, even predation. What is required to address short-termism, then, is a very public process of discerning productive capital from non-productively placed assets. Doing so may be described as the 'socialization' of capital. Government's role in capital socialization must proceed cautiously, however, within efficiency parameters established empirically within the field of competitiveness research, pioneered by Harvard Business School's Michael Porter and others.

4.2 US short-termism: evidence

Short-termism has been rampant in the US economy for at least two decades. It lies at the polar extreme from what has been called 'patient capital', that is, finance committed by investors in long-term relationships with companies undertaking long-term strategies compatible with the nurturance of product innovation. A good working definition of short-termism, compatible with a recent seminal article on the subject by John R. Graham et al., is: an excessive focus on short-term quarterly earnings, combined with a lack of attention to strategy, fundamentals, and conventional approaches to the creation of long-term value (Graham et al., 2005).

Enron is a classic – but by no means isolated – example of egregious, short-term behaviour to which the Sarbanes-Oxley (2002) Corporate Anti-fraud Act responded. The 'players' are well known within the American financial press; among them: Kenneth Lay and Jeffrey Skilling of Enron, Bernard Ebbers and Scott Sullivan of WorldCom, and Adelphia's John Rigas, to name but a few of the most visible ones. Eighty-year-old Rigas was sentenced in 2004 to 15 years in prison – effectively a life sentence – for defrauding the cable company of more than $100 million. He and his family also hid more than $2 billion in debt while lying about Adelphia's condition (Searcey and Yuan, 2005).

Detailing the Enron debacle in *The Smartest Guys in the Room*, Peter Elkind and Bethany McLean (the *Fortune Magazine* investigative reporter who broke the Enron story in March 2001) described the fraud-laced, short-term corporate culture that sobered the American public, at least momentarily (Elkind and McLean, 2004). Stunning, particularly, is the esteem in which Enron was held; it regularly topped 'respected business leaders' surveys from among the nation's most innovative large companies.

An excerpt taken from Enron's 1998 Annual Report describes its self-perceived culture as anchored in respect (We treat others as we would like to be treated ourselves.... Ruthlessness, callousness, and arrogance don't belong here); integrity (We work... openly, honestly, and sincerely); communication and excellence. Regarding innovation, *Fortune Magazine* named Enron 'Most Innovative Company' six years in a row (Elkind and McLean, 2004, p. 23). Kenneth Lay and Jeffrey Skilling were widely viewed as business visionaries. Lay died of a heart attack in mid-2006, prior to his sentencing on multiple counts of fraud; three months later Skilling was sentenced to 24 years in prison.

It was Jeffrey Skilling who designed the financial innovations having the effect of freeing Enron's natural gas business from its physical constraints (Elkind and McLean, 2004, p. 37). The Gas Bank, as it was known, actually freed Enron from holding production and transportation assets. The company could simply own a portfolio of contracts that allowed it to control resources it needed, instead of viewing contractual commitments to deliver natural gas as something that required a physical pipeline. As in the case of the Gas Bank, 'the realms of production and finance were thought to have become interconnected – even seamless – through financial innovation'.[1]

Profitable operation of the Gas Bank required, among other Skilling financial innovations, the legitimatization and adoption of 'market-to-market' accounting. Skilling insisted on market-to-market as a condition of coming to Enron from his lucrative consultancy with McKinsey and Company. Under conventional GAAP (generally accepted accounting principles), an asset's value on the company's books reflects initial assumptions over its life, even if the underlying economics eventually change. Revenues and profits flowing in from contracts – including forward contracts for the delivery of natural gas – are booked as they come in the door, quarter by quarter, year by year. But in market-to-market, at Enron, estimated changes in asset value (the present value of the estimated income stream over an asset's life) were capitalized in the present and therefore showed up immediately as current income for Enron. This allowed Skilling – in the first year of a multi-year contract – to book the entire estimated appreciation over the asset's life (Elkind and McLean, 2004, p. 39).

Until Enron, the principal use of market-to-market had been among traders. At Enron, however, it became a tool to 'game' the natural gas business – producers, consumers, investors and employees, all.

During the Enron Congressional oversight hearings in 2002, Representative W. J. Tauzin, Chairman of the House Committee on Energy and Commerce, observed:

> ... we are continuing to explore the policy implications of the Enron collapse ... Enron filed false and misleading accounting statements that were not in accordance with GAAP ... Enron's management engaged in self-dealing that amounted to theft of shareholder assets and the directors and auditors who were supposed to supervise the corporation failed to prevent this theft ... [Further] ... Enron had a tangled web of off-balance sheet partnerships serving to hide [from the market] billions of dollars in debt and other liabilities. (House Committee on Energy and Commerce, 2002)

Yet, in spite of Sarbanes-Oxley and an increasing national focus on short-term, fraud-oriented financial practices, these continue in the US economy seemingly unabated. Among recent contributors to the literature on short-termism is John Bogle, applied ethicist, highly visible scion of American capitalism, and founder of the venerable family of indexed mutual funds, Vanguard. In his 2005 book, *The Battle for the Soul of Capitalism*, Bogle minced no words about the consequences of this widespread behaviour (Bogle, 2005). He cites Edward Gibbon's 1838 classic in his first paragraph, and draws the parallel between the contemporary United States and *The Decline and Fall of the Roman Empire*.

Bogle describes the contemporary 'bread' of material goods acquisition and self-indulgence, and the 'circuses' of wealth as a measure of self-worth, combined with the need for personal recognition. In short, Bogle takes a position against greed. Like the Romans, he says, America's vision of greatness is fading and Gibbon's history reminds us that no nation can take its greatness for granted. Bogle particularly decries soaring executive compensation linked by incentives to quarterly earnings guidance, and he presents a series of recommendations for 'the reform of this faltering system' (Bogle, 2005, p. 241). In many ways Bogle's reform proposals mirror the conclusions of a nationally prominent commission report on industrial productivity, published 14 years earlier (Dertouzos et al., 1989).

Other visible writers have made similar appeals, also with brilliance. Among them is a 2005 contribution by Alfred Rappaport to *Financial Analysts Journal* (Rappaport, 2005). Regarding the increasing intensity of short-termism, he observes the average holding period for stocks until the mid-1960s was about seven years. Currently, however, the average holding period in professionally managed funds has decreased to less than one year; therefore annual portfolio turnover is greater than 100 per cent. Contemporary investment managers can create an informationally efficient market, he says, without simultaneously making it highly efficient, allocatively. Rappaport's analysis highlights the distinction between real product innovation, on the one hand, and financial innovation on the other. He concludes,'... the market provides no free lunch, but it does provide occasional early bird specials for the most skillful investors'. Among the 'specials', of course, are other manifestations of short-termism including trades on 'inside information'.

The influential Association of Chartered Financial Analysts (CFA)/Centre for Financial Market Integrity, the Business Roundtable/Institute for Corporate Ethics, and the Conference Board/Corporate-Investor Summit Series each weighed in recently with common diagnoses of short-termism and essentially common prescriptions for containing it. The titles: *Asset Manager Code of Professional Conduct* (Chartered Financial Analysts, 2005), *Breaking the Short-Term Cycle* (Business Round Table Institute for Corporate Ethics, 2006), and *Revisiting Stock Market Short-Termism* (Conference Board, 2006) are descriptive of increasing public concern and continuing intractability of the problem, almost two decades following its initial, thorough diagnosis. By most accounts, short-termism grows more pernicious over time. A serious observer may conclude, then, that inaction is indicative of the inability of the private sector to reform itself, a lack of will to do so, or both. Nor is comprehensive government oversight likely to succeed, because the status quo does not distinguish, effectively, differences between productive and non-productive asset-holding strategies.

Certainly Enron's collapse was a stinging wake-up call to look carefully at egregious financial practices including the use of pro forma financial statements and market-to-market accounting, among others. However, Carrie Johnson and Ben White, writing for *The Washington Post*, describe how years after the collapse of Enron and the subsequent Sarbanes-Oxley legislative response, chemistry for other financial disasters remains (Johnson and White, 2006). Companies continue to face enormous pressure to meet short-term financial targets, they point out, creating a powerful motive for the perpetuation of accounting fraud. Wall Street analysts continue to seek guidance about company revenues, quarter after quarter. And businesses continue to pay steep prices for missing targets by even a penny or two per share.

Worse still, according to former Securities and Exchange Commission (SEC) Chairman Harvey Pitt, government regulatory efforts may have an unintended consequence. 'Many shareholders ... have been led to believe [these reforms] have cured all the problems and we're home free,' says Pitt. 'Unfortunately, that's a prescription for disaster.' As an example, commodities-trading firm Refco collapsed in 2005 amid an accounting scandal, just two months after its initial public offering, led by prominent investment bank Goldman Sachs (Johnson and White, 2006).

At the Enron trial, the Justice Department's case included accusations that earnings-per-share were 'tweaked' by Jeffrey Skilling. In the practice of 'smoothing', executives manipulate financial statements to obtain consistent earnings performance from quarter to quarter. For instance, prosecutors charged in one incident that Skilling pulled a penny 'out of thin air' to avoid falling short of Wall Street stock analysts' expectations. On cross examination, Enron's former investor relations chief testified that on 17 January 2000 – the day before the company was to report quarterly results – he informed Skilling and others that Wall Street's consensus earnings forecast had risen by a penny, from 30 cents a share to 31. Mark Koenig testified that by the next morning the earnings announcement had been altered to reflect the 31-cent analysts' expectation. Koenig conceded he was not sure where the extra penny came from, but said he believed it was manufactured to meet analysts' expectations (McWilliams and Scannell, 2006).

Another illustration of short-termism is occurring among so-called 'big box' retailers including executives at home improvement giant Home Depot, who are 'pushing back' against Wall Street analysts. Recently Home Depot announced it was doing away with a component of earnings guidance, same-store sales 'comps'. Comps, in this case, compare revenues from stores open more than one year with revenues from those same stores in the prior year. According to *Wall Street Journal*'s Jesse Eisinger, same-store comps are the most carefully monitored measure of American retailing performance, but executives believe, because of them, investors overreact frequently, triggering stock price volatility (Eisinger, 2006). In public efforts to protect itself by controlling information flows, Home Depot may be exacerbating short-termism, however. Thus, analysts may now be pressured to speculate about the magnitude of comp numbers, adding more fuel to the bonfire.

Economists are also uncovering evidence that some companies pass up profitable opportunities if they perceive such an investment will have a negative short-term impact on earnings. One of the most disturbing recent empirical studies, by the National Bureau of Economic Research, reported that when researchers asked a series of questions about the importance of quarterly earnings to 400-plus US executives, a surprising 78 per cent acknowledged they would sacrifice long-term value in order to generate same-period earnings. Managers also reported they work to maintain predictability in earnings and other financial disclosures, a revelation that is consistent with the short-termism hypothesis (Graham et al., 2005).

In the wake of the Enron-era accounting scandals, observes co-author Campbell Harvey of Duke University, companies have been less willing to use 'accounting shenanigans' as he calls them, such as those described here, to meet earnings projections. Now, instead, '…they are doing things that affect the real operations of the company, like postponing research and development (R&D)', which is '…the stuff that creates real value in the long term' (Nocera, 2006). In other words, in order to manage stock prices in a post-Enron, Sarbanes-Oxley world in which the 'degrees of freedom' to manipulate earnings and other data are reduced, CEOs are now jiggling R&D expenditures. 'If you can't make your short-term profit target, so this logic goes, then make it up by "drubbing" R & D', seems to be the favoured strategy. This, of course, destabilizes innovation and with it, long-term productivity growth.

4.3 US short-termism: historical perspective

In 1989 the MIT-appointed Commission on Industrial Productivity issued its path-breaking report, *Made in America: Regaining the Productive Edge*, which in part was motivated by the 1980s pre-eminence of the Japanese economy over the American system. Written by Michael Dertouzos, Richard Lester and Nobel laureate Robert Solow, the report 'indicted' American industry on multiple counts, including short-termism, or as they said, 'the quest for a high return on investment that runs counter to the realities of world trade' (Dertouzos et al., 1989).

Then one year later, Michael Porter published *The Competitive Advantage of Nations*, a seminal study in which he laid out a theory of competitive strategy spanning several academic fields (Porter, 1990). In it Porter defined the evolution of national economies through 'stages', from factor-driven, to investment-driven, to innovation-driven and potentially, then, into decline associated with the ultimate, 'wealth-driven' stage. It is in this final stage in which powerful firms and individuals may be motivated and able to insulate their wealth positions by influencing government policy (Porter, 1990, p. 556).

There is a sobering reality in the wealth-driven stage, observes Porter, that wealth has been achieved in prior periods rather than being created dynamically. The challenge to rentiers in this stage, then, is not so much to create wealth, but to maintain it and pass it on intergenerationally.

This is because, most importantly, the motivations of investors, managers and individuals in this stage begin to shift away from one another in ways that undermine productive innovation.

Then, two years later in 1992, the *Harvard Business Review* published Porter's 'Capital Disadvantage: America's Failing Capital Investment System' (Porter, 1992). The study, co-sponsored by the US Council on Competitiveness and based upon the work of a distinguished, interdisciplinary, Porter-led team, reported on short-term investments and the time horizons of US corporations.

The conclusion? The US economic system advances the goals of shareholders interested in near-term appreciation, even at the expense of the longer-term performance of the US economy. In this somewhat 'zero-sum' environment, competing groups of managers, investors, analysts, etc. become trapped with others in the rationality of their own ends, to the exclusion of other classes of economic 'players'. Thus, they fail to benefit from the synergy which tended to bind prior generations of their predecessors in a 'non-zero-sum' environment in which 'the whole tended to exceed the sum of the individual parts'. This conclusion is particularly chilling because it reveals a potential systemic flaw in American-style corporate capitalism; that is, the pursuit of self-interest may not articulate fully with the attainment of the common good.

The Porter team identified reform proposals including: (1) the slowing of so-called 'stock market churning' through a transactions tax on securities trading volumes (often described as the 'Tobin Tax', after Yale University's James Tobin); (2) improving alignment between investors' and managers' goals; (3) strengthening the proxy system or increasing the number of outside directors, and (4) increasing the use of stock options in management compensation (as has surely occurred more recently), but with restrictions also placed on managers' abilities to exercise options, unless they simultaneously meet specific long-term targets. A sample of some of the Commission's more 'specific' proposals is contained in note 2.

Of particular interest, here, is the reality that short-termism has gained attention at the highest levels of policy and scholarship and now approaches a two-decade run. Unfortunately – the Sarbanes-Oxley Act notwithstanding – the American system of corporate capitalism appears unable to correct itself in the face of the ratcheting upward of both violations and also proposals for addressing them. One wonders, then, does the US lack the will to fix the problem? Or more fundamentally, do alternatives for a viable fix even exist within the US status quo?

This chapter offers a bold hypothesis, that short-termism is a creature not only of financial manipulations, but also of doctrinal malfunction influencing the character of American-style corporate capitalism, particularly. Short-termism, unfortunately, is embedded in received neoclassical doctrine. Therefore, aggressive efforts to solve the problem through self-policing – even the expansion of government oversight – are doomed to failure, it is asserted. Similar to the first glitch of doctrinaire capitalism – Say's Law – this second glitch or profit lacuna has recently become a problem because 'the way the world works' has changed, even as the foundational doctrines remain static.

4.4 Discussion[3]

The first glitch or malfunction of doctrinaire capitalism was revealed with the Great Depression, seven decades ago. Simply put, the received paradigm did not recognize disequilibrium and therefore did not admit long-term unemployment and economic gluts. John Maynard Keynes's analysis and prescription, simply put, had the effect of repealing Say's Law that the act of providing this year's output simultaneously provides a flow of income from producers to households, enabling the purchase of that output level. This recognition that disequilibrium may occur in actual economies patterned after doctrinal principles led to a revised economic security role for government, implemented in response to the Great Depression of the 1930s.

It was Cambridge economist Joan Robinson, particularly, who called attention to the second doctrinal glitch: imprecision between the respective productivities of capital and finance (Robinson, 1971). Since the neoclassical system lacks an inherent theory of what the rate of profit on capital should be, there emerge potential and actual confusions between capitalists and rentiers; that is, between the pursuit of profit and the pursuit of rent. Keynes called for the euthanasia of rentiers. It is argued, then, that the contemporary manifestation of Keynes's rentiers are so-called pseudo-capitalists who pursue goals and utilize strategies generally incompatible with the common good. These are people and institutions such as Enron that 'game' the system for their own self-serving ends, while largely going unapprehended. Why is this so?

Across two centuries, the conventional wisdom of doctrinal capitalism has asserted that the pursuit of self-interest articulates universally with the attainment of the common good when executed through the institutions of private property and minimally regulated markets. For pre-industrial or classical societies, the central problem within this frame pertained to wealth creation, and for industrial or neoclassical societies, the central problem pertained to the explanation of cost or value. However, wealth or cost-driven perspectives neither describe nor adequately address the pre-eminent, contemporary problem of post-industrial societies – particularly the United States – which is how to publicly choose among alternate futures; that is, how to determine what the future shall be in pluralistic society, how access to future outcomes shall be distributed, and the rewards that shall flow to entrepreneurs who facilitate such outcomes.

A crucial related problem is how to police non-productive behaviours in which self-interested action does not articulate with the attainment of the common good. Such action may occur when pseudo-capitalists and other rentiers enable themselves to be compensated as legitimate capitalists, but without proffering the requisite industry, skill and risk-taking anticipated by society in exchange for the preferred treatment it confers upon them.

The very nature of capital is in transition as industrial outcomes proceed towards an emerging post-industrial outcome. Some characteristics are satiation of basic goods and services, technological acceleration, a widening dispersion between the 'haves' and the 'have nots', a deteriorating environment, and the qualitative pursuit of lifestyle, among others. Fierce competition exists in post-industrial society for scarce 'lifestyle goods' such as strategically located real estate, fashionable cars and clothes, exotic vacations, entertainment and sporting events, among others.

In a qualitative sense capital in this emergent reality must now be conceived as whatever is necessary to move from one set of outcomes or potentialities, to the future attainment of some other set. Since capital investment is about the cultivation of capability, it is rooted in some vision or scenario of the future. Indeed, contemporary capital ultimately exists relative to the outcomes that wealthy, democratic societies choose to create.

Rather than speaking of laissez-faire capitalism or mixed capitalism, then, the emergent post-industrial society might better be characterized as relative capitalism. That is, contemporary capital requires specification – not absolutely – but relatively, relative to some vision it is engaged in creating. Otherwise, it becomes merely a rentier asset engaged in extending the status quo.

To accommodate this emergent reality, Say's Law must be rejected anew, and with it the presumption that Newtonian order prevails in real economies modelled upon doctrinaire principles. With the admissibility of disequilibrium in which planned saving may diverge from planned investment, rent-seeking behaviours become admissible theoretically. In a practical sense, pseudo-capitalists are characterized as those who may save or control the saving of others, but fail to expand social output to potential, and may actually contribute to disinvestment.

The roots of pseudo-capitalism are innocent enough; indeed, they are staples of conventional instruction in management education and applied policy analysis. Consider the so-called least cost rule of the marginal productivity theory of resource demand. 'A firm is producing a specific output with the least-cost combination of resources when the last dollar spent on each resource yields the same marginal product' (McConnell and Brue, 2005). A corollary is that one dollar's worth of any asset should receive the same high rate of return as one dollar's worth of any other asset. Otherwise, the financial asset holder may reduce opportunity cost and therefore increase portfolio yield by replacing lower-yielding assets with higher ones.

But this ostensibly rational dictum at the level of the individual or organization too often becomes reified and prescribed as a path of rational action for entire societies. For instance, consider how rent-seeking behaviours may be revealed within a contemporary economic environment in which trillions of dollars are traded electronically each day in global financial markets. In such an environment a product may be designed, financed, produced and marketed in entirely different countries. Here, industrial capitalists who choose to do so may morph conveniently into post-industrial, short-term-oriented, rent-seeking pseudo-capitalists.

This transformation from capitalist to pseudo-capitalist, described by Sawyer, begins with the dictum of marginal productivity theory that one dollar's worth of any asset P should receive the same rate of return as one dollar's worth of any other asset C, where P and C denote pseudo-capitalist and capitalist assets, respectively. For instance, we begin with the capitalist holding C and the pseudo-capitalist holding P. Soon the capitalist may be motivated to morph into a pseudo-capitalist because he recognizes the opportunity cost of holding C is the extra skill, industry and risk-taking that s/he perceives to now go uncompensated. The unmitigated pursuit of self-interest may ultimately lead all capitalists to act as pseudo-capitalists. To the extent this occurs, then societal outcomes are reduced by the value of the extra contributions indicative of the difference in social product between the capitalist class and the rentier class (Sawyer, 2005).

Pseudo-capitalist behaviours are not likely to raise opposition from the ranks of either conventionally trained policy makers or the academic keepers of the paradigm who sustain them. Buried beneath its surface, however, are embedded assumptions that may have appeared benign in the industrial age but have profound ethical implications in the post-industrial age. Among these is the assertion that the pursuit of self-interested action universally articulates with the attainment of the common good. While this may have been a reasonable generalization of authors writing a century ago about the Industrial Revolution, now it too often describes merely an ideological position rather than reality. In part this has transpired because the way in which the world works has changed, while the way in which we think about how the world works has remained remarkably static in the face of a global revolution in the way business is conducted.

Because it denies disequilibrium, the methodological invocation of general equilibrium is also blind to hoarding behaviours in which capitalists plan to invest less than households plan to transmit to them in household saving, in the simple two-sector formulation. This occurs because entrepreneurs may be motivated to act as pseudo-capitalists, but the political–economic system is blind to their manipulations and continues to reward them as legitimate capitalists. Contemporary hoarding takes many non-productive forms including accounting frauds and other egregious, rent-seeking behaviours, described here as short-termism.

To address short-termism effectively, a new perspective is required that realigns conventional economic thinking about the way in which the world works, with the way it actually works. Included therein must be a dramatically revised role for government, to limit short-termism by vetting capital productivity. Since doctrinaire capitalism does not stipulate a disequilibrium rate of return on capital assets – distinguishable from rentier assets – government must enter to set the rate of profit on capital, thereby curbing short-term-oriented hoarding behaviours.[4] Government's pre-eminent new role within contemporary relative capitalism, then, becomes the coalescence of a vision of the future with enumeration of the capital assets necessary to attain it, then the creation of an incentive structure to bring the vision into being.

4.5 Conclusion

It is argued here that short-termism is a creature not just of the innovative manipulation of finance, but of a doctrinal second glitch that in disequilibrium fails to distinguish the rewards of rentiers from the rewards of capitalists. Consequently, short-termism should be viewed less as an aberration and more as an expected manifestation of self-interested, finance-oriented behaviours in sophisticated, contemporary economies. These behaviours are resistant to public oversight because they are inculcated into the conventional wisdom. Paradigmatic roots undergird the manner in which students of business and economics are educated and socialized. As mature capitalist economies have segued from industrial to post-industrial, short-term-oriented hoarding behaviours by pseudo-capitalists have become more flagrant and also more costly to the long-term well-being of investors, specifically, and society generally. Short-termism and other contemporary rentier behaviours can be resolved only by first resolving the second glitch of doctrinaire capitalism. Practically, this necessitates the socialization of capital through setting its rate of profit, including the structuring of incentives, positive and negative.

4.6 Postscript: (appropriate) government role

Renaud Bellais of the University of Littoral implicitly provides a broad sketch of an implied solution to short-termism – the 'socialization of capital' – as an appropriate response to a theoretical void. Published in the *Journal of Post Keynesian Economics* in 2004, the Bellais article entitled 'Post-Keynesian Theory, Technology Policy, and Long-Term Growth' clarifies that research and innovation are not addressed as central features of conventional post-Keynesian models, in part because these models integrate technological progress through the process of capital accumulation (Bellais, 2004). Ultimately, he observes, post-Keynesian (distinct from neoclassical) models cannot integrate – but from a short-term perspective – uncertainty, risk assessment, and investment decisions because the sole source of uncertainty is unexpected variation in the level of aggregate demand; the crucial determinant of investment within the model. An important implication is that the business sector may be unable to support 'blue sky' basic research adequately. This is because 'the innovation effect is embodied in the investment function....' [Practically]...'When firms give up funding fundamental and applied research', innovation and therefore investment are compromised; consequently, the state – concludes Bellais – 'should organize a complementary policy'.

Regarding government role, however, Michael Porter offers a crucial qualification (Porter, 1990, Ch. 12). Says Porter, 'Government's proper role in enhancing national advantage is the reverse of what is often supposed'... many of the ways in which government tries to 'help' can actually hurt a nation's firms in the long run [for example, government policies such as support for subsidies, domestic mergers, high levels of industrial cooperation, guaranteed government demand, and artificial devaluation of the currency, among others]. Rather, the appropriate government role, he observes, is that of a 'pusher and challenger'. Thus, argues Porter, 'There is a vital role for government-induced pressure – even adversity – in the process of creating national competitive advantage.'

One of government's most essential roles is 'signalling'. A good example of signalling is the campaign launched by the Japanese government, beginning in the 1960s, to elevate national attention regarding product quality, ultimately to overcome the post-Second World War stigma of cheap Japanese export goods. The most powerful levers available to government for influencing national competitive advantage, observed Porter, are *slow-acting* ones such as creating advanced factors, encouraging domestic rivalry, shaping national priorities and influencing demand sophistication. Conversely, he concludes, 'The quick, easy roles of government (subsidy, protection, macroeconomic management) are either insufficient or counterproductive.' In sum, 'Government should not overstate or overplay its role in national competitive advantage. If it does, it will create an economy of dependent, backward-looking, and ultimately unsuccessful firms.'

What is attempted here, then, is a sketch of a paradigm-appropriate government role in mature, evolving capitalist societies, motivated particularly by the call to overcome short-termism. This sketch seeks to steer a course between the macro-oriented 'socialization-of-investment-role' for government, advocated by Bellais, on the one hand, and potentially productive and counterproductive 'micro' roles for government, on the other, identified by Michael Porter and others within the field of competitiveness research. While advocating the socialization of capital, government actions should proceed cautiously; that is, capital socialization should be circumscribed by empirically known principles of competitiveness.

Notes

1   Jeffrey Skilling believed the highly regulated – then deregulated – natural gas industry that operated precipitously through boom and bust cycles could get out of its predicament by creating a Gas Bank. By the late 1980s, 75 per cent of natural gas transactions in the US occurred on the spot market during a frantic

few days at the end of each month. The problem for producers, particularly, was inherent uncertainty. It was risky for pipelines to contract to deliver a steady supply of gas to industrial customers at a price that ensured a profit. Enron Gas Marketing was created to provide a steady supply over extended periods to customers willing to pay Enron a hefty risk premium. Eventually it contracted with customers not even connected to Enron's pipeline. According to McLean and Elkind, it was the first serious effort to diminish the level of risk for all players in natural gas transactions. Enron, the bank, captured profits between its 'buy' and its 'sell' prices. Through hedging, Enron conceived of a balanced portfolio of contracts – that is, contracts to sell at a given price were exactly offset by contracts to buy at the same price. Thus, regardless of spot market prices, Enron would already have fixed its risk and locked in its profit margin. Customers were enthusiastic, but suppliers were reluctant, sensing the opportunity cost of selling their gas low, then watching an appreciating resource create wealth for a broker. To attract sellers, Skilling offered to cash out suppliers from long-term contracts by loaning them the net present value of the contracts up front. In effect, say the authors, Enron's business plan effectively freed the natural gas industry from the physical qualities of the resource it supplied. Instead of seeing a commitment to deliver gas as something requiring a pipeline, instead Enron conceived it as a financial commitment. It was a whole new way of conceptualizing the business, one that required less capital, at least theoretically, and also produced more stable pricing and more flexibility for customers.

2   Some of Porter's specific proposals are:
   - Eliminate restrictions on joint ownership of debt and equity.
   - Reduce the extent of subsidies for investment in real estate.
   - Modify accounting rules so that earnings better reflect corporate performance.
   - Expand public disclosure to reduce the cost of assessing true corporate value.
   - Allow disclosure of 'insider' information to significant long-term owners, under rules that bar trading on it.
   - Loosen restrictions on institutional board membership.
   - Encourage board representation by significant customers, suppliers, financial advisers, employees and community representatives.
   - Codify long-term shareholder value rather than current stock price as the appropriate corporate goal.
   - Extend tax preferences only to those stock options and stock purchase plans with restrictions on selling.
   - Provide investment incentives for R&D and training.

3   This section draws heavily upon James E. Sawyer: 'Reframing Capitalism' in *John Kenneth Galbraith and the Future of Economics*, Blandine Laperche and Dimitri Uzunidis (eds) (London: Palgrave Macmillan, 2005); and also *Why Reaganomics and Keynesian Economics Failed* (London and New York: Macmillan and St. Martin's, 1988).

4   Indicative economic planning, as the term may be used, for instance, in the economies of Austria and Japan, is roughly proximate to the concept of setting the rate of profit on capital.

References

Bellais, R., 'Post Keynesian Theory, Technology Policy, and Long-Term Growth', *Journal of Post Keynesian Economics*, 26 (2004), 419–40.

Bogle, J. C., *The Battle for the Soul of Capitalism* (New Haven and London: Yale University Press, 2005).

Business Round Table/Institute for Corporate Ethics, *Breaking the Short-Term Cycle* Washington, DC: (2006).

Chartered Financial Analysts (CFA), Center for Financial Market Integrity, *Asset Manager Code of Professional Conduct* (Charlottesville, Va, 2005).

Conference Board, Corporate/Investor Summit Series, *Revisiting Stock Market Short-Termism* (New York:2006).

Dertouzos, M., R. Lester and R. Solow. *Made in America: Regaining the Competitive Edge* (Cambridge: MIT, 1989).

Eisinger, J., 'Long and Short: Retailers Discount Sales Stats', *Wall Street Journal*, 7 June 2006.

Elkind, P. and B. McLean, *The Smartest Guys in the Room* (New York: Penguin, 2004).

Graham, J. R., H. R. Campbell and S. Rajgopal, 'The Economic Implications of Corporate Financial Reporting', *Journal of Accounting and Economics*, 40 (2005), 3–73.

House Committee on Energy and Commerce, Press Release, 14 February 2002: http://energycommerce.house.gov/107/news/02142002_495.htm

Johnson, C. and B. White, 'Conditions Still Ripe for another Enron', *The Washington Post*, 12 February 2006.

McConnell, C. R. and S. L. Brae, *Economics: Principles, Problems, and Policies*, 16th edn (Columbus, Ohio: McGraw-Hill, 2005).

McWilliams, G. and K. Scannell, 'Profit Tweaking May Lose Favor after Enron Trial', *Wall Street Journal*, 16 February 2006.

Nocera, J., 'A Defense of Short-Termism', *The New York Times*, 29 July 2006.

Porter, M., *The Competitive Advantage of Nations* (New York: Free Press, 1990).

Porter, M., 'Capital Disadvantage: America's Failing Capital Investment System', *Harvard Business Review* (Reprint 92508) (September–October 1992), 65–82.

Rappaport, A., 'The Economics of Short-Term Performance Obsession', *Financial Analysts Journal*, 61 (2005), 65–79.

Robinson, J., *Economic Heresies* (New York: Basic Books, 1971).

Sawyer, J. E., *Why Reaganomics and Keynesian Economics Failed* (London and New York: Macmillan and St. Martin's, 1988).

Sawyer, J. E., 'Reframing Capitalism', in Blandine Laperche and Dimitri Uzunidis (eds), *John Kenneth Galbraith and the Future of Economics* (London: Palgrave Macmillan, 2005).

Searcey, D. and L. Yuan, 'Adelphia's John Rigas Gets 15 Years…', *Wall Street Journal*, 21 June 2005.

5
Finance, State and Entrepreneurship in the Contemporary Economy

Sophie Boutillier

Entrepreneurs have been at centre of economists' concerns and public policies since the beginning of the 1980s. This fact is relatively new. Since the end of the Second World War, the paradigm of the big company has prevailed. A paradigm (Kuhn, 1983) is defined as a set of interrogations, assumptions and responses starting from which a given reality becomes understandable. However, a paradigm also gives a structure to the scientific community which uses it as a reference. To question the theoretical model of reference is also to shed some doubts concerning the authority of the scientific community. On this aspect, we cannot fail to recall the acid remarks of J. Robinson, when she exposed the unrealistic nature of the production function, insufficient to explain economic reality, but absolutely necessary to make an academic career (Robinson, 1984). The paradigm of the big enterprise did not develop without clashes and conflicts. It progressively asserted itself to oppose the paradigm of a pure and perfect competition ignoring the existence of enterprises of different sizes (Walras, 1988).

The years of growth that followed the Second World War were marked by a phenomenon of industrial concentration and the evolution of managerial capitalism (Section 5.1). Economy was directed by a 'technostructure' (Galbraith, 1968) and, in particular, by managers being salaried workers (Chandler, 1977). Entrepreneurs, as founders-owners-managers of firms, seemed to belong to an age that had gone, to the heroic period to which J.A. Schumpeter often refers. The big company imposed itself, and together with it mass production and salaried employment (see Table 5.1). In 1968, W.J. Baumol (1968) wrote in a famous paper that the entrepreneur had disappeared from the economic literature.

Table 5.1 From one paradigm to another

Paradigms	Paradigm of pure and perfect competition	Paradigm of the big company	Paradigm of the entrepreneur. Updating of Schumpeter's theory about the entrepreneur
Periods	End of the nineteenth century until the Second World War	Years 1950–70	Since the 1980s
Basic principles and concepts	Market atomicity Product homogeneity Information transparency Free entry into the market Free movement of production factors The entrepreneur is an intermediary between the production factors market and the market of goods and services	Economies of scale Market organization Uncompleted monopolistic competition Partial equilibrium Outsourcing Public policies/regulation The manager replaces the entrepreneur Big enterprises are governed by a 'technostructure'	Decline/restructuration of the big enterprise Small enterprises are more dynamic than bigger ones in terms of innovation and create more jobs Outsourcing policies of big enterprises (network enterprise) and new investment opportunities
Main authors	L. Walras	J.A. Schumpeter Mark II J.K. Galbraith A. Chandler	M. Casson G. Gilder S. Shane

But how did the paradigm of the entrepreneur emerge? The entrepreneur, as a concept, is reappearing in economics because of the presence of positive factors that contribute to create a propitious environment for the creation of enterprises. We hold the idea according to which the economic, social and political environment facilitates the development of specific economic behaviours, as for example entrepreneurial behaviour (Boutillier and Uzunidis, 1995, 1999, 2006; Julien, 2005). According to the OECD (2005), the emergence of entrepreneurship is related to the rank it holds in the scale of values and to the intensity of incentives and support it receives. However, the beginning of the 1980s was marked by a whole set of major transformations that consecrate a sort of rupture from the previous period: (1) public policies targeted at deregulation/privatization; (2) development of financial markets; (3) emergence of new information and communication technologies and biotechnologies; (4) decline of public expenses; (5) augmentation of mass unemployment and implementation of public policies devoted to its reduction through incentives for business creation. It is in this new context that the entrepreneurial function evolved. Theses changes accelerated starting from the 1990s, in particular after the fall of the Berlin wall and the collapse of the major socialist economies. These events announced the end of history (Fukuyama, 1993) or rather the end of a story. Capitalism has become the only form of social organization, combining the power of the market in economic regulation and that of democracy in political regulation. That period was also the starting point of the development of various works devoted to different forms of capitalism. As rightly recalled by B. Amable (2005), studies made on the different performances of capitalist economies date from the 1970s. Nevertheless, since the beginning of the 1990s, and the institutional upheavals that have marked that period, economists have been compelled to develop a new research field dealing with the characteristics of capitalism. This research fields concentrates on institutions (North, 1990), in order to isolate those institutions, values, behaviours, etc. that facilitate economic growth and those which, to the contrary, contribute to its deceleration (Section 5.2).

In the first years of the twenty-first century the economy of industrialized countries is undergoing major transformations at the scientific and technological levels. If one refers to Schumpeter's theory about entrepreneurship, this situation lays a fertile ground for innovation and for business creation (Langlois, 1987; Perroux, 1965; Heertje, 2006), a process that fuels the ascending phase of an economic cycle (see Chapter 2, written by D. Uzunidis on this point). As did J.-B. Say in his time, we may also assert that the entrepreneur plays the full role of intermediary between the scientist who produces knowledge and the worker who applies it to industry. While historians and economists used to talk about a first, second and third industrial revolution (Freeman and Louçã, 2001), it seems that, for a few years, the term 'industrial revolution' has gone astray and has been replaced by other expressions such as knowledge economy or knowledge society. After the steam engine, the motor vehicle, jet aircraft, the new technological area is characterized by different opportunities of miniaturization (e.g. electronic components) and by an acceleration of information dissemination. This scientific and technological restlessness is inscribed in a specific context characterized by the development of financial markets whose actors are constantly on the watch for short-term profitable investments. To use Keynes's metaphor, venture capital firms and investment funds are on the lookout for [the best opportunity] (Keynes, 1982; Galbraith, 1992).

In this context, the entrepreneur is no longer heroic, but rather socialized (Boutillier et al., 2006). He is stuck between two logics: that of the big enterprise that restructures and outsources all or a part of its activities; and that of the state striving to promote the creation of new businesses, on the one hand to fight against unemployment and, on the other, to foster the development of innovations seen in the Schumpeterian meaning of the term (product, process, organization). The concept of the socialized entrepreneur must be distinguished from the collective entrepreneur or even, from the entrepreneurial corporation (Hagedoorn, 1996) that characterizes the managerial enterprise: in fact, the socialized entrepreneur may be defined in the first place by his macroeconomic function (job creation, innovation, outsourcing of the productive and service activities of big companies).

5.1 The building of the big enterprise's paradigm

5.1.1 The heroic entrepreneur, engine of industrial capitalism

The major interest of Schumpeter's analysis of the entrepreneur lies in its multidisciplinary combination of three disciplines: economy, sociology and history. He was not satisfied by describing and deepening the economic theory. That is why he also proceeded to a rather subtle sociohistorical analysis by putting the stress on two basic elements: the private ownership of the means of production and the recognition of business success. The necessity to reiterate the importance of the social and cultural context in which the individual is anchored is a consequence of that. Schumpeter argued that entrepreneurs are chosen from among the 'new men', those who question established habits and routines. How can we explain the acceleration of economic growth starting from the eighteenth century? Many economists give as an explanation the development of entrepreneurial activity (Murphy et al., 2006). To be an entrepreneur is neither a profession, nor, above all and in general, a long-lasting situation (Schumpeter, 1979). The capitalist economy is subjected to a permanent process of 'creative destruction' characterized by the renewal of investment opportunities resulting from technological progress. At the beginning of the twentieth century, the popular classes accessed consumer goods that the powerful monarchs of past centuries would never have imagined.

In Schumpeter's *Theory of Economic Evolution* (Schumpeter, 1935), the entrepreneur is the economic agent achieving new combinations of production factors. He is the economic agent who promotes a new economic dynamic and introduces technological progress into the economy. In this case, Schumpeter adopted the paradigm of a pure and perfect competition that does not make it possible to give an interpretation of technological progress, the dynamic 'growth-decline' or the existence of firms of different sizes. Finally, we can only understand Schumpeter's theory on the entrepreneur (and, simultaneously on innovation) by placing it in the history of innovation theory (Hagedoorn, 1996). Schumpeter questioned the paradigm of pure and perfect competition, while asserting that Walras was the greatest of all economists (Schumpeter, 1983; Langlois, 1987). Is not the expression 'heroic entrepreneur' slightly exaggerated? If one reads the biographies of some entrepreneurs, the answer is immediately negative. One of the many biographers of Louis Renault begins his Chapter 4 as follows: What is the builder of an empire? Do his morphology and figure reflect power and audacity? (Dingli, 2000, p. 191). *The (heroic) Schumpeterian entrepreneur* (Freeman, 1982), quoting Schumpeter Mark I, is the economic agent who achieves new combinations of production factors. Five combinations must be taken into account:

1. Manufacturing of a new good in particular unfamiliar to consumers' circles or endowed with a new quality.
2. Introduction of a new production process that is almost unknown in the specific industrial branch; it is not imperative that it is based on a new scientific discovery and it may also be found in the new commercial processes applied to a commodity.
3. Opening of a new outlet, a market in which the specific industrial branch of a specific country has not yet been penetrated, irrespective of the previous existence of that market.

4. Acquisition of a new source of raw materials or semi-finished products; again it does not matter whether this source has to be created or already exists, has been taken into consideration or considered inaccessible.
5. Formation of a new organization, for example creation of a monopolistic situation or sudden emergence of a monopoly. The heroic entrepreneur is the entrepreneur who creates a new industry, similar to what happened at the end of the nineteenth century (movies or electricity) (Mangolte, 2006).

In his ultimate book entitled *Capitalism, Socialism and Democracy*, published in 1942, Schumpeter was largely pessimistic about the future of capitalism. He took up the analysis of Marx and to the question 'Will capitalism survive?', he answered 'No'. It was not because he foresaw a political revolution, the victory of the proletariat over the wealthy, but because the development of capitalism led, according to him, to the disappearance of competition. While Schumpeter gave greater place to the economic aspects of the issue (economies of scale and extension of the firm's size), we cannot ignore, on the other side, the social and political context in which he wrote (Russian revolution, world wars, etc.) (Swedberg, 1991). Companies were becoming bigger and bigger. In addition, these were powerful organizations, bureaucratic enterprises. Schumpeter insisted on the following idea: the entrepreneur is being replaced by an organization. Entrepreneurs are no longer responsible for innovative activities, which are now performed by teams composed of expert members who have no direct link with the market or the consumer. In such conditions, business, the creation of enterprises ceases to be an adventure, an instrument of self-achievement. In a world dominated by big companies, the market seems frozen. Only the domains pertaining to culture, sports, education, arts, etc. will constitute in the future an engine of self-achievement.

The vanishing of the Schumpeterian entrepreneur is a metaphor used to analyse the development of managerial capitalism, the evolution of big enterprises. The entrepreneur constitutes another metaphor with which to understand the dynamic of technological progress. At the beginning of the twentieth century, this idea was shared by a considerable number of researchers, economists and sociologists (Veblen, Hilferding, Berle and Means, Burnham, etc.). For example, M. Weber (1989, 2003) explained that civil society was being militarized. Enterprises operated like armies; their organization was streamlined. H. Fayol gave a definition of the actual principles of contemporary management. F. Taylor dealt with the scientific organization of labour and clocks the workers' gestures; H. Ford introduced the conveyor-belt system. The dividing line between production work and design activities became clearer. 'White collars' besieged the very few big companies controlling the American market (Wright Mills, 1966, 1969). Big enterprises had become pyramids concentrating all the manufacturing operations in one building having the size of a small town: new materials entered through one door and the finished car exited through the door situated at the opposite end (Sennett, 2006, pp. 33–4). This integration went very far. For example, Ford was the owner of steel and coal mines (Berger, 2005, p. 88). In spite of the vanishing of the heroic entrepreneur, the economic growth that followed the end of the war was extremely rapid and without any precedent in human history. Was Schumpeter wrong? (Heertje, 2006).

5.1.2 Post-war economic growth: shareholder and manager vs entrepreneur

The paradigm of the big enterprise was reinforced after the Second World War while the state became a primordial economic actor both as regards supply (support policy, assistance to job seekers, families, pensioners, etc.) and demand (social policy, assistance to job seekers, pensioners, etc.). This increasing role of the state was obviously the counterpart of the ever-growing investments needed by industry, but it also revealed the continuation of a trend which started during the Second World War and linked to the industrial and technological effort of that time (armaments, scientific programmes, supplies, etc.). It is also the counterpart of the context of the cold war that followed the signature of the Yalta agreements consecrating a new division of the world between two parties that were relatively well defined – capitalist world and socialist world – corresponding a priori to the two different economic logics: market vs state. However, J.K. Galbraith considered that the common points between the two economic systems largely prevailed over their differences (Galbraith, 1968): the role of the state was quite important in the United States (military–industrial complex) and big American enterprises made their plans according to their activity in the market (disappearance of competition to the benefit of oligopolistic markets).

Managerial capitalism was not born after the Second World War. Its origin is much older as demonstrated by the works of the economists T. Veblen (1971) and R. Hilferding (1970) in the United States and Europe. The development of the stock exchange and of financial markets by the end of the nineteenth century strongly boosted industrial capitalism because new industries (railways, iron and steel, chemicals, etc.) needed increasing investments that could not possibly be financed by one individual only. But it is undoubtedly the study performed by Berle and Means, published in 1932, which introduced this question in the academic realm, contributing to the questioning of the paradigm of pure and perfect competition. Almost simultaneously, E. Chamberlain (discriminatory monopoly) and J. Robinson (imperfect competition) quite actively participated in the elaboration of the paradigm of the big enterprise. But, from the 1920s onwards, the works of A. Marshall had contributed to the erosion of the Walrasian paradigm (externality, partial equilibrium, short/long term, etc.). At the beginning of the 1940s, J. Schumpeter (1935, 1979) went back on what he had written in his youth. He explained that the entrepreneur of the heroic times will disappear and leave the place to big enterprises. Innovation becomes the business of a team of experts who cease to be directly related to the market.

In the 1960s, J. K. Galbraith (1968) pursued Schumpeter's analysis of managerial capitalism and demonstrated that the economy of capitalist industrialized countries did not fit with the paradigm of pure and perfect competition. Six distinctive elements emerged:

1. Domination of a handful of big enterprises whose ownership is split between a myriad of shareholders, a plethora of small owners of the enterprise (is it still possible to talk about 'private property'? In the middle of the nineteenth century, K. Marx wrote that such a phenomenon celebrates the destruction of private property);
2. Presence of a considerable number of very small firms, however rather marginal as regards the creation of wealth;
3. Disappearance of the entrepreneur replaced by a division between the owners of capital (shareholders) and capital management (managers); the 'technostructure'. The manager (salaried and non-owner of the enterprise) maximizes his own interest (growth of the firm's size; technical virtuosity);
4. Development of planning tools in order to minimize the uncertainty resulting from the very functioning of the market. The economists who are the founding fathers of the theory of the entrepreneur (R. Cantillon, J.-B. Say, J. A. Schumpeter, F. Knight; Esposito and Zumello, 2003; Formaini, 2001) argued in fact that the entrepreneur is the economic agent who bears the risk resulting from the uncertainty inherent in the functioning of the market. The combination between the questioning of the assumption of market atomicity and the idea according to which the (big) enterprise organizes the

market unavoidably leads to the questioning of the very existence of ... capitalism!
5. Presence of a plethora of entrepreneurs (point 2) who are at the same time the owners and managers of the assets they valorize. However, these small entrepreneurs do not operate in a market characterized by pure and perfect competition, but in markets dominated by big enterprises.
6. Similarities between capitalism and socialism: both in the United States and the Soviet Union, the development of the big enterprise is characterized by the expansion of a huge bureaucracy, related to technological and not political considerations. The management of a large-size company requires a whole series of expertise. As ironically stated by Galbraith, it is not a Schumpeterian entrepreneur who sent men in the space, but a bureaucracy; and this irrespective of the political regime which characterized the USA or the Soviet Union.

Now that the stage has been set, what were during the period 1950–70 the engines of economic growth and of big enterprises?

1. War economy: for more than five years the United States and Europe have substantially invested in industry in order to foster the production of military material, but also strongly contributed to the financing of scientific research. The war gave a strong impulse to the development of several industries: aeronautics, motor vehicles, chemicals, telecommunications, sectors that were largely responsible for the growth of the post-war economy. One should also mention the cold war which stimulated the arms race as well as the political and economic competition between the two blocks (Kennedy, 1991; Galbraith, 1968).
2. The development of an abundant supply of standardized consumer goods (cars, electric appliances, etc.) led to the development of mass consumption. The car industry developed quickly after 1950; cars which were still luxury items in the 1930s became basic consumer products. Big automotive companies extracted from their drawers some sketches already made before the war with a view to the production of popular vehicles, as typified by Model T (Ford) and Hitler's Volkswagen, without forgetting the 4HP manufactured by Renault and the 2HP of Citroën (Boutillier and Uzunidis, 1999, 2006). Air transport and mass tourism quickly developed; households acquired telephones and electric appliances. Towns grew and hypermarkets appeared at their periphery; banks multiplied their consumer credit

supplies (mainly for cars and housing); households became indebted (Verret, 2007).
3. Big enterprises made investments (development of Taylorism and Fordism); salaried employment increased (workers in industry, employees in the service sector), guaranteeing better remuneration and extending ancillary social services (social security, unemployment insurance, family allowances, pensions) (Report of the Society of Nations, 1943; Beveridge, 1944; UN Report, 1949). This period was also marked by the development of a state bureaucracy characterized by the implementation of national accountancy instruments with a view to better plan economic development (Massé, 1965; Sauvy, 1970).
4. The role of the state increased under the pressure of conflicts (1939–45; cold war) (Peacock and Wiseman, 1961) in order to sustain both demand (minimum wage, family allowances, education, health, retirement, etc.) and supply (enterprises and public investments; policy of 'national champions'; development of education, research, etc.).
5. Financial markets were rather slow to develop. Financing of the economy was essentially done through banks that were in majority public institutions (indebtedness economy).

The economic, social and political context of the period 1950–70 was far from being favourable to heroic entrepreneurs. The hegemony of the big managerial enterprise seemed to be without any limit: IBM, Coca-Cola, General Motors, Ford became the symbols of a new mode of both consumption and production. To show its extraordinary know-how in the field of marketing, Coca-Cola reinvented Father Christmas and adorned him with its colours, white and red.

5.2 The socialized entrepreneur and the new capitalism

5.2.1 The turning point of the 1980s, a context favourable to the socialized entrepreneur

Starting from the end of the 1970s, major changes came to question again that dynamic of economic growth:

1. Crisis of the Bretton Woods Agreements and the international monetary system: devaluation of the dollar and loss of competitiveness of the American economy.

2. Liberalization of the economy (contestable markets theory) and development of the financial markets (the privatization of the economy releases capital in huge quantities: new investment opportunities emerge; development of investment funds and pension funds; the ageing of the population and the withdrawal of the social state from the financing of pensions stimulated their development) (Aglietta and Rebérioux, 2004). The major problem was to identify new investment opportunities in a context of slow economic growth (Galbraith, 1992). The activity of banks evolved: business banks became internationalized. Owing to the increasing sophistication of financial instruments such as the leveraged buyout (LBO), investors were now capable of creating or dismantling companies without any possibility of opposition from their management. Due to their overwhelming position investors no longer demanded long-term results but short-term ones (Sennet, 2006). Capital became impatient (Harrisson and Blustone, 1990). While in 1965 American pension funds held shares over a period of 46 months, in 2002 a fair part of the portfolio of these institutional actors was renewed on average every 3.8 months. The exchange of shares on the basis of market prices got the better of traditional performance indicators like the price–earnings ratio: this was in particular the case in the 1990s when title values rocketed in companies that made no profit (Sennett, 2006, pp. 39–40).
3. Development of new information and communication technologies and biotechnologies (Castells, 1990): scientific and technological progress generated new investment opportunities. The beginning of the 1980s was marked by the creation of new types of enterprises in the microelectronics field (Apple, Microsoft, etc.). The 'garage' mythology expanded and the legend of the entrepreneur prevailed (Boutillier and Uzunidis, 1999). As in the early days of capitalism, an idea that was already considered outdated was revived and propagated: young brilliant inventors may create their enterprise and become rich. This is the rebirth of the heroic entrepreneur. However, one tends to forget that the knowledge they used to succeed is the result of military research and of the efforts made by big companies (computers, microprocessors, etc.).
4. Crisis of the welfare state: G. Gilder (1985) argued that the social state generates poverty because it encourages too many people to rely on social services instead of looking for a job (since the 1970s, the 'public choice' school or the theory of bureaucracy have strongly criticized Keynesianism). Only the entrepreneur is capable to fight against poverty. R. Laffer questioned again the redistributive power

of taxation. The major question was no longer to create sufficient demand in order to boost consumption, but to channel it towards the creation of enterprises; in short, to resuscitate the old-fashioned entrepreneur described by Galbraith.
5. New public policy: since the beginning of the 2000s, state policies have had the obligation to create a favourable frame for the development of business. The World Bank (2006) defines a synthetic indicator to evaluate the performance of the state in this field. That composite indicator includes the following factors: creation of enterprises, granting of licences, recruitment of wage-earners, credit supply, protection of investors, payment of taxes, international trade, execution of contracts and closing down of enterprises. A world classification is established based on the performance of the state at that level. Earlier, Keynesian economists had explained that the fundamental role of the state was to sustain demand and create markets. On the other hand, the economists of endogenous growth theories explain that the state has a major role to play in order to sustain the supply and support enterprises to innovate; in its turn, innovation generates wealth and employment. The stress is put on public investments and on the development of human capital (increase in the number of researchers in enterprises and universities), as well as on the commercialization of research (policy to facilitate the filing of patents and enlargement of the domain of patentability – e.g. patentability of the living parts). A new field of economic analysis is developing: the knowledge economy which considers that the production of goods and services depends on the achievement of increasing knowledge-intensive scientific and technological progress, at the highest level of expertise (Foray, 2000).
6. Increase of mass unemployment and growing insecurity of salaried employees (multiplication of the number of term contracts instead of permanent contracts, development of temporary work): is this the end of labour (Rifkin, 1995)? Of a salaried society (Castel, 1995)? Through an appropriate public policy, the state tries to facilitate the transition from the situation of job seeker to that of entrepreneur, or from wage-earner to entrepreneur; in short, to introduce more flexibility in the labour market (Boutillier, 2006, 2007).
7. The big managerial enterprise with its pyramidal architecture (Sennett, 2006) is no longer adapted and is compelled to change: the network enterprise appears on the scene; its organization is decentralized (to benefit from new information and communication technologies) and remains managerial. Its assets are distributed between a relatively

important group of major shareholders. By definition, managers lead the enterprise in spite of the growing pressure of the financial markets. Neither managers nor shareholders are attached to the enterprise employing them. A managerial elite migrates from one enterprise to another according to mergers and acquisitions. It seems that its objective is no longer the growth of the firm, as was the case during the Fordist period. By making the best possible use of the liquidity of the stock markets, the purpose of that elite is to extract from the enterprises maximum cash for its own benefit. Since business banks are interested in triggering a maximum number of security transactions, it is not surprising that mergers and acquisitions explode at the time of stock exchange euphoria (Aglietta and Rebérioux, 2004, p. 343; Laperche et al., 2006).
8. Seven characteristics define the networked firm (Castells, 1998): (i) organization around a process, not a task; (ii) flat hierarchy; (iii) team management; (iv) measure of the collective results; (v) maximization of contacts with suppliers and customers; (vi) information; (vii) education and continuous training of wage-earners at all levels (Castells, 1998, Book I, p. 221). Thus, new information and communication technologies promoted the development of subcontracting. There are many examples. Apple's iPod is manufactured by a myriad of subcontractors. Enterprises-contractors are thus free from the constraints related to lead time and investment costs (Berger, 2006, pp. 100–1).
9. As regards the number of salaried workers/employees, the size of enterprises has also been reduced: in America and Great Britain, the majority of enterprises have less than 3000 employees; many of them have only a local character or are family businesses; some are craft enterprises similar to small building companies. They may very well operate in the form of small bureaucratic pyramids (Sennett, 2006, pp. 37–45).
10. Since the beginning of the 1990s, entrepreneurship has become an academic discipline taught in universities. Awareness programmes targeted at the youth are also elaborated upon (Audet, 2001; Gasse and Tremblay, 2002; Riverin and Jean, 2006).

5.2.2 Theory of the socialized entrepreneur in a knowledge economy dominated by finance

A socialized entrepreneur is an entrepreneur sitting at the interface between two logics:

1. The logic of the big industrial and financial enterprise that seeks to stimulate the creation of enterprises in order to test new markets.

2. The logic of the state that seeks by these means to fight against unemployment and promote innovation. The entrepreneur is defined on the basis of his resource potential.

We define the entrepreneur as the founder, manager and owner of at least a part of the enterprise. In such conditions, he may also be an innovator (innovation can also be at the origin of the creation of the firm, Schumpeter said); however, unemployment may as well be at the origin of his decision (Casson, 1991). Nevertheless, he always remains the economic agent who bears the risk (Cantillon, 1755; Knight, 1921) since he is in every case, the main financial backer of his enterprise, together with his relatives.

The concept of 'resource potential' is the result of a treble interrogation anchored in the determination to go beyond the concepts of *homo oeconomicus* and the firm:

1. Formation of a network of social relations for the economic agent: according to neoclassical theory, an individual makes his decisions on the basis of a cost/benefit analysis. Therefore, he is a *homo oeconomicus*. This idea was later enriched (bounded rationality, anticipation, etc.) by some economists. The works performed on social capital by some sociologists since the beginning of the 1980s (Bourdieu, 1985; Coleman, 1988; Putman, 1995; Adler and Kwon, 2002) has induced many researchers to elaborate on that concept in order to show how the entrepreneur mobilizes his social relations network to achieve his project (Taylor et al., 2004). The support provided by relatives is often viewed as a determining factor (Jack and Anderson, 2002; Davidson and Honig, 2003). The social relations network also constitutes a means of collecting information (legislation, subsidies, technologies, etc.) necessary for the launching of the project (Suire, 2004; Corbett, 2002; Craig and Lindsay, 2001) or to identify investment opportunities. Sometimes, results are disconcerting since, quite often, future entrepreneurs do not find what they are looking for. Opportunities discovered by entrepreneurs are frequently related to their prior information and expertise (Shane, 2004).
2. The enterprise as a complex of resources: at the end of the 1950s, E. Penrose (1959) defined the enterprise as a set of productive resources structured in a managed framework. The function of the firm is to acquire and organize its material, intangible and human resources with the purpose of selling with a profit on the goods and services market. Penrose's idea was to open the black box represented by the

Table 5.2 Resource potential of the founder of an enterprise: elements for a general definition

Resource potential	Major characteristics
Knowledge	Tacit and various types of knowledge acquired in the family context Scientific and technological knowledge acquired at school Knowledge acquired during our relations with third parties (family, professional activity)
Financial resources	Own spending Affective inputs: parents, relatives Bank credit Institutional financial aid (e.g. direct assistance from the state) Financial inputs brought in by another entrepreneur
Social relations	Informal relations (family, friends, neighbours, colleagues, etc.) Formal relations (state, banks, other enterprises, research centres, etc.)

enterprise in neoclassical theory and restricted to a mere productive function.
3. The entrepreneur as distributor of resources between markets (Walras, 1988); however, he may also be considered as an input contributing to the acceleration of economic growth (Baumol, 1993).

By using the concept of 'resource potential' (Table 5.2), we relocate the entrepreneur and his enterprise in the industrial logic and system. The resource potential is split up in the following way:

1. A set of financial resources including all the effective financial resources (own spending, family assets, heritage) or potential (access to credit, subsidies, various aids).
2. A set of knowledge including all the entrepreneur's knowledge whether they are certified by a diploma or result from professional experience: technological, organizational, economic knowledge, etc.
3. A set of social relations: personal, family or professional relations that the entrepreneur may mobilize in order to fulfil his project. Two social relations networks may be distinguished: on the one hand, a network of institutional relations (relations with public institutions, enterprises, etc.), on the other, a network of informal relations (relations with relatives, family or friends, neighbours, working relations, etc.). In our example, these two networks develop interdependently. Thus, it is through the information given by a friend that we learn

about the existence of a specific type of financing. However, the individual's social background plays a fundamental role because it largely determines the network of friendly relations.

The three components of the entrepreneur's resource potential are determined by the place he holds in the social or family organization chart – in spite of the increasing socialization of the economy. These elements assume a fundamental role. The family gives a taste to start a business; at the same time, it is a source of financing. With the support of the family, the functions exerted by the entrepreneur draw their logic from public policies targeted at the dampening of the consequences of the crisis (employment and innovation policies) and from strategies aiming at the productive and financial reorganization of big enterprises.

Therefore, this capacity of the entrepreneur results from the variety and richness of the resource potential he has himself constituted. In its turn, the composition of that resource potential depends on factors that are external to the enterprise and entrepreneur. In particular, public policies of assistance for the creation of businesses (to stimulate innovation and/or to fight against unemployment) will largely determine the financial resources to which the entrepreneur will be authorized to have access in order to create his enterprise and ensure its survival. During the 1990s and the 2000s, the multiplication of 'management shops' and other associations of assistance to business creation, has become particularly significant. Many job seekers, sometimes having exhausted their benefits, become entrepreneurs and are successful. The economic and social organization has several dimensions and therefore several effects. The general level of development of knowledge and technology in the society will have an impact both on the knowledge acquired and assembled by the entrepreneur (on the basis of his education and the competences of the members of his team; activities related to economic and information watch) and the technological level of his activity. The nature of the financial system (e.g. ease or difficulty of going public, bankers' degree of 'conservatism', level of development of venture capital, etc.) influences both the capacity of an individual to become an entrepreneur and the capacity of an enterprise to more or less accelerate its development. As demonstrated earlier, the degree of concentration in the market, for example the presence of big enterprises, also plays a considerable role in the dynamic of creation of small enterprises and in their type of activity (in particular, subcontracting). Finally, it is necessary to underline the policy led by big enterprises with a view to innovate, either by their own means (R&D budget) or by implementing different types of partnership

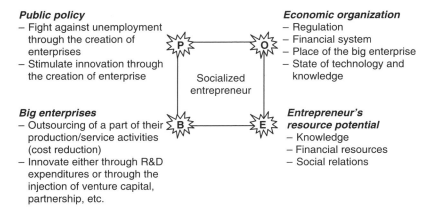

Figure 5.1 The socialized entrepreneur
Source: Boutillier et al. (2006)

including the injection of venture capital. Finally, the presence and nature of the links between the 'POBE' factors (Public policy – Economic organization – Big enterprises – Entrepreneur's resource potential) lead us to relocate the entrepreneur in his economic, social, political, technological and scientific context. This organic square provides a way to analyse the creation of enterprises at the scale of a specific economy (Figure 5.1).

At the beginning of this twenty-first century, almost all enterprises are SMEs, not pyramidal enterprises. This conclusion seems surprising when we look at the analyses made by many economists during the first half of the twentieth century: unable to adapt to modernity, the small enterprise and entrepreneur were doomed to disappear. How can we understand this paradox when the entrepreneur's main function, according to economic analysis, was to bear the risk inherent in the market. Does the disappearance of the entrepreneur mean that competition will vanish? Most probably (Schumpeter, 1975; Galbraith, 1968). The decision taken by big enterprises, at the beginning of this twenty-first century, to outsource huge segments of their activities (maintenance, engineering, production of semi-finished goods, etc.) provides many opportunities to create new enterprises. Even more as we observe that, already for many years, public policies targeted at the struggle against unemployment have been oriented towards the creation of businesses. Job seekers must have the opportunity to become entrepreneurs and big enterprises must find new scientific and technological resources in order to innovate (see Table 5.3).

Table 5.3 Big enterprises and entrepreneurs during the second half of the twentieth century

Periods	Place of big enterprises	Organization of labour and production	Place and role of the entrepreneur	Form of recruitment	Financing of the economy	Role of the state
During years of economic growth	Development of big enterprises (managerial capitalism)	Assembly chain Fordism Taylorism Rigid organization	Entrepreneur = employer = authority	Mass wage earning Term contract Mass unemployment	Indebtedness (important role of banks) Public financing	Highly developed social state
Since the 1980s	Reorganization of big enterprises (network enterprise)	NTIC Robotization of production and services Flexible organization	Entrepreneur = innovator = creator	Increasing precariousness of salaried employment Term contract Mass unemployment	Development of financial markets	Privatization of big public enterprises Questioning of the social state creation of a favourable context for the development of business

Bibliography

Adler, P. S. and S.-W. Kwon, 'Social Capital: Prospects for a New Concept', *Academy of Management Review*, 27 (1) (2002), 17–40.
Aglietta, M. and A. Rebérioux, *Les dérives du capitalisme financier* (Paris: Albin Michel, 2004).
Amable, B., *Les cinq capitalismes* (Paris: Seuil, 2005).
Audet, J., 'Une étude des aspirations entrepreneuriales d'étudiants universitaires québécois: seront-ils des entrepreneurs demain?', Document de travail, Institut de Recherche sur les PME, Université du Québec à Trois-Rivières (2001).
Baumol, J. W., 'Entrepreneurship in Economic Theory', *The American Economic Journal*, 58 (2) (1968), 64–71.
Baumol, J. W., *Entrepreneurship, Management and the Structure of Payoffs* (Cambridge, Mass.: The MIT Press, 1993).
Berger, S., *How we Compete: what Companies around the World are Doing to Make it in Today's Global Economy* (New York: Doubleday Broadway, 2006).
Berle, A. A. and G. G. Means, *The Modern Corporation and Private Property* (New York: The Macmillan Company, 1932).
Beveridge, W., *Full Employment in a Free Society* (London: G. Allen & Unwin, 1944).
Borges, C., G. Simard and L. J. Filion, 'Entreprendre au Québec, c'est capital! Résultats de recherche sur la création d'entreprises', *Cahiers de recherche de la chaire d'entrepreneuriat Rogers-J.-A. Bombardier* No. 3 (2005).
Bourdieu, P., 'The Forms of Capital', in J. G. Richardson (ed.), *Handbook of Theory and Research for the Sociology of Education* (Westport: Greenwood, 1985), pp. 241–58.
Boutillier, S., 'The End of Capitalism: J. K. Galbraith versus K. Marx and J. A. Schumpeter', in B. Laperche, J. K. Galbraith and D. Uzunidis (eds), *Innovation, Evolution and Economic Change: New Ideas in the Traditions of Galbraith* (Cheltenham: Coll. New Directions in Modern Economics, Edward Elgar, 2006), pp. 53–70.
Boutillier, S., 'Politique publique et création d'entreprise. Une analyse contemporaine du capitalisme aménagé de Keynes' (Dunkirk: Cahiers du Lab. RII, ULCO, No. 154, http://rii.univ-littoral.fr. 2007).
Boutillier, S. and D. Uzunidis, *L'entrepreneur, une analyse socio-économique* (Paris: Economica, 1995).
Boutillier, S. and D. Uzunidis, *La légende de l'entrepreneur* (Paris: Syros, 1999).
Boutillier, S. and D. Uzunidis, *L'aventure des entrepreneurs* (Paris: Studyrama, 2006).
Boutillier, S., B. Laperche and D. Uzunidis, 2006, 'The Entrepreneur's "Potential" and Innovation in Contemporary Economies', Conference on Rethinking Entrepreneurialism in the University Context, 10–11 November, Saint Petersburg, Russia.
Burnham, J., *The Managerial Revolution* (New York: John Day Co, 1941).
Cantillon, R., *Essai sur la nature du commerce en général* (1755) (Paris: Institut National d'Etudes Démographiques, 1952).
Casson, M., *L'entrepreneur* (Paris: Economica, 1991).
Castel, R., *La métamorphose de la question sociale* (Paris: Gallimard, 1995).
Castells, M., *La société en réseaux* (Paris: Fayard, 1998).
Chamberlin, E., *Theory of Monopolistic Competition: a Reorientation of the Theory of Value* (Cambridge, Mass.: Havard University Press, 1933).

Chandler, A., *La main visible des managers* (Paris: Economica, 1977).
Coleman, J. S., 'Social Capital in the Creation of Human Capital', *American Journal of Sociology*, 94 (1988), 95–120.
Corbett, A. C., 'Recognizing High-tech Opportunities: a Learning and Cognitive Approach', *Proceedings of the Twenty-Second Annual Entrepreneurship Research* (Babson Park, Mass.: Arthur M. Blank Center for Entrepreneurship, Babson College, 2002), pp. 49–60.
Craig, J. and N. Lindsay, 'Quantifying "Gut Feeling" in the Opportunity Recognition Process', *Proceedings of the Twenty-First Annual Entrepreneurship Research* (Babson Park, Mass.: Arthur M. Blank Center for Entrepreneurship, Babson College, 2001), 124–36.
Davidson, P. and B. Honig, 'The Role of Social and Human Capital among Nascent Entrepreneurs', *Journal of Business Venturing*, 18 (3) (2003), 301–31.
Dingli, L., *Louis Renault* (Paris: Flammarion, 2000).
Esposito, M.-C. and C. Zumello (eds), *L'entrepreneur et la dynamique économique* (Paris: Economica, 2003).
Foray, D., *Economie de la connaissance* (Paris: Coll. Repères, La Découverte, 2000).
Formaini, R. L., 'The Engine of Capitalism Process: Entrepreneurs in Economic Theory', *Economic and Financial Review*, 4 (2001), 2–11.
Freeman, C., *The Economics of Industrial Innovation* (Cambridge, Mass.: The MIT Press, 1982).
Freeman, C. and F. Louça, *As Time goes: From the Industrial Revolutions to the Information Revolution* (Oxford: Oxford University Press, 2001).
Fukuyama, F., *La fin de l'histoire et le dernier homme* (Paris: Flammarion, 1993).
Galbraith, J. K., *Le nouvel état industriel* (Paris: Gallimard, 1968).
Galbraith, J. K., *Brève histoire de l'euphorie financière* (Paris: Seuil, 1992).
Gasse, Y. and M. Tremblay, *L'entrepreneuriat à l'université Laval: intérêts, intention, prévalence et besoins des étudiants*, Rapport d'analyse (Laval: Centre d'entrepreneuriat et de PME, Université de Laval, 2002).
Gilder, G., *L'esprit d'entreprise* (Paris: Fayard, 1985).
Hagedoorn, J., 'Innovation and Entrepreneurship: Schumpeter Revisited', *Industrial and Corporate Change*, 5 (3) (1996), 883–96.
Harrisson, B. and B. Blustone, *The Great U-Turn. Corporate Restructuring and the Polarizing of America* (New York: Basic Books, 1990).
Heertje, A., *Schumpeter on the Economics of Innovation and the Development of Capitalism* (Cheltenham: Edward Elgar, 2006).
Hilferding, R., *Le capitalisme financier* (Paris: Editions de minuit, 1970).
Jack, S. L. and A. R. Anderson, 'The Effects of Embeddedness on Entrepreneurial Process', *Journal of Business Venturing*, 17 (5) (2002), 478–88.
Julien, P. A., *Entrepreneuriat régional et économie de la connaissance. Une métaphore des romans policiers* (Montreal: Presses de l'université du Québec, 2005).
Kennedy, P., *Naissance et déclin des grandes puissances* (Paris: Payot, 1991).
Keynes, J. M., *Théorie générale de l'emploi, de l'intérêt et de la monnaie* (Paris: Payot, 1982).
Kickul, J. and L. K. Gundry, 'Pursuing Technological Innovation: the Role of Entrepreneurial Posture', in *Frontiers of Entrepreneurship Research* (Babson Park, Mass.: Arthur M. Blank Center for Entrepreneurship, Babson College, 2000), pp. 200–10.

Knight, F., *Risk, Uncertainty and Profit* (Chicago: Chicago University Press, 1921).
Kuhn, T., *La structure des révolutions scientifiques* (Paris: Flammarion, 1983).
Langlois, R. N., 'Schumpeter and the Obsolescence of the Entrepreneur', The History of Economics Society annual meeting (Boston: 21 June 1987).
Laperche, B., J. Galbraith and D. Uzunidis (eds), *Innovation, Evolution and Economic Change. New Ideas in the Tradition of Galbraith* (Cheltenham: Edward Elgar, 2006).
Mangolte, F., 'Naissance de l'industrie cinématographique. Les brevets aux Etats-Unis et en Europe (1895–1908)', *Annales HSS*, 5 (2006), 1123–45.
Marczewski, J., *Comptabilité nationale* (Paris: Dalloz, 1965).
Massé, P., *Le Plan ou l'anti-hasard* (Paris: Gallimard, 1965).
Murphy, P. J., J. Liao and H. P. Welsch, 'A Conceptual History of Entrepreneurial Thought', *Journal of Management History*, 12 (1) (2006), 12–34.
North, D., *Institutions, Institutional Change and Economic Performance* (Cambridge: Cambridge University Press, 1990).
Peacock, A. T. and J. Wiseman, *The Growth of Public Expenditure in the United Kingdom* (Princeton: Princeton University Press, 1961).
OECD, *Perspectives de l'OCDE sur les PME et l'entrepreneuriat* (Paris: OECD, 2005).
Perroux, F., *La Pensée economique de Joseph Schumpeter. Les dynamiques du capitalisme* (Geneva: Dalloz, 1965).
Putman, R. D., 'Bowling Alone: America's Declining Social Capital', *Journal of Democracy*, 6 (1) (1995), 65–78.
Rifkin, J., *La fin du travail* (Paris: La découverte, 1995).
Riverin, N. and N. Jean, *L'entrepreneuriat chez les jeunes du Québec: état de la situation (2004)* (Montreal: Chaire d'entrepreneuriat Rogers-J.-A. Bombardier, 2006).
Robinson, J., *The Economics of Imperfect Competition* (1933) (London: Macmillan, 1969).
Robinson, J., *Contributions à l'économie contemporaine* (Paris: Economica, 1984).
Sauvy, A., *La prévision économique* (Paris: Coll. Que sais-je?, PUF, 1970).
Say, J.-B., *Traité d'économie politique* (Paris: Crapelet, 1803).
Schumpeter, J. A., *Théorie de l'évolution économique* (Paris: Dalloz, 1935).
Schumpeter, J. A., *Capitalisme, Socialisme et démocratie* (Paris: Payot, 1979).
Schumpeter, J. A., *Histoire de l'analyse économique* (Paris: Gallimard, 1983).
Sennett, R., *La culture du nouveau capitalisme* (Paris: Albin Michel, 2006).
Shane, S., *A General Theory of Entrepreneurship: the Individual–Opportunity Nexus* (Cheltenham: E. Elgar, 2004).
Sing, R. P., G. E. Hills, R. C. Hybels and G. T. Lumpkin, 'Opportunity Recognition through Social Network Characteristics of Entrepreneurs', in *Frontiers of Entrepreneurship Research* (Babson Park, Mass.: Arthur M. Blank Center for Entrepreneurship, Babson College, 1999), pp. 228–41.
Society of Nations, *Le passage de l'économie de guerre à l'économie de paix* (Geneva: SDN, 1943).
Suire, R., 'Des réseaux de l'entrepreneur aux ressorts du créatif: quelles stratégies pour les territoires?', *Revue internationale des PME*, 17 (2) (2004), 121–42.
Swedberg, R. (ed.), *Joseph A. Schumpeter: the Economics and Sociology of Capitalism* (Princeton: Princeton University Press, 1991).
Taylor, D. W., O. Jones and K. Boles, 'Building Social Capital through Action Learning: an Insight into the Entrepreneur', *Education Learning*, 1 (2004), 226–35.

UN Report, *Le maintien du plein emploi* (New York: UN, 1949).
Veblen, T., *Les ingénieurs et le capitalisme* (Paris: Editions des archives contemporaines, 1971).
Verret M., *Le travail ouvrier* (Paris: A. Colin, 2007).
Walras, L., *Eléments d'économie pure* (Paris: Economica, 1988).
Weber, M., *L'éthique protestante et le capitalisme* (Paris: Pocket, 1989).
Weber, M., *Economie et société* (Paris: Pocket, 2003).
World Bank, *Doing Business* (Washington, DC: World Bank, 2006).
Wright Mills, C., *Les cols blancs* (Paris: Ed. François Maspéro, 1966).
Wright Mills, C., *L'élite au pouvoir* (Paris: Ed. François Maspéro, 1969).

6
The Political Economy of R&D in a Global Financial Context[1]

Jerry Courvisanos

6.1 Introduction

Expenditure on research and development (R&D) forms the basis of incremental innovation for medium-to-large corporations in all advanced economies. Successful innovation out of R&D provides a high rate of return to the economy. This is because, although the vast majority of projects fail to materialize any tangible results, these failures contribute significantly to knowledge accumulation in the innovation process. However, the high risk and fundamental uncertainty arising out of R&D raise concerns about funding such activity. The new global financial system places this concern into a much sharper context.

The funding concerns in the R&D process need to be analysed at the macroeconomic level, and a political economy (PE) approach is adopted to conduct this enquiry. This approach is based on a dilemma created by the dual role of R&D. On the one hand, R&D aims to generate knowledge and promote the development of new ideas into products, processes and services (called 'innovations') that drive economic growth. On the other hand, it entrenches monopoly power within corporations that undertake R&D and then patent or copyright the results. The former role encourages the extension of the frontiers into new knowledge, while the latter tries to limit knowledge by marginal incremental marketing-based improvements that have more to do with planned obsolescence. Understanding this dual role of R&D in the context of the need to finance such risky and uncertain R&D is crucial. How can new knowledge be encouraged and financed when it is safer (and more short-term profitable) to fund marginal extensions of current knowledge?

The twin roles of R&D are reflected in the work of the first major researcher on innovation, Joseph Schumpeter. In his first analysis in 1911, Schumpeter (1934) identified the entrepreneurial process in terms of the small capitalist who drives new ideas into the marketplace while destroying old products and processes (which Schumpeter called 'creative destruction' in his 1942 book), and this seemed to be consistent with the form of capitalism observed by economists through the nineteenth century. The innovative activity is seen to be exogenous to the firm (especially the characteristics of the entrepreneur), in what has been referred to as Schumpeter Mark I. There is no official R&D undertaken under Mark I, as small entrepreneurs develop, test and market their ideas in the process of producing and selling the idea. Financing Mark I R&D, in essence, is part of the whole difficult process of raising funds by small entrepreneurs (see Bhidé, 2000).

By the early 1940s, Schumpeter – in his Mark II form – recognized institutionalization of R&D as sustaining the monopoly power of corporations to the point that he was concerned that this process would see the end of the entrepreneur and R&D becoming a purely bureaucratic activity (Schumpeter, 1942). This raised the spectre of 'creative accumulation' from minor incremental innovative activity that is endogenous to the corporation. Galbraith (1967) developed further the Mark II analysis by identifying R&D as the endogenous innovation process which attempts to manage the problems of both market and technical uncertainty that emerge out of any new products and processes. In this way, the large corporation's 'technostructure' sets up strategic planning and related investment commitments in the context of secrecy and intellectual property rights from R&D-generated innovations. Financing Mark II R&D by large corporations is based on past profit outcomes and willingness to plough such funds back into uncertain future-based projects with expectations of significant private returns.

This chapter is a survey of the extant academic literature on R&D from the particular focus of its twin roles and the implications that has for financing R&D. The next section aims to first examine the role of R&D in the innovation process through the innovation dilemma between its role as a knowledge-generating process and the entrenched power that such knowledge creates as economic growth is achieved. The impact of financing R&D is then identified in the context of a short-term perspective emerging as monopoly power is arrogated to a few large companies with vast resources and huge market shares that undermines new knowledge R&D. This leads to a set of rationales for public financing of R&D. Finally, the complete R&D process and its financing are placed within the global financial system that has become established early in the twenty-first century.

6.2 Role of R&D in the innovation process

R&D is a complete process 'whereby new and improved products, processes, materials, and services are developed and transferred to a plant and/or market. Typically, this process is represented in the firm by a number of formally organized laboratories, departments, groups, teams and functions ... most easily recognized [that] ... involve scientists and engineers' (Burgelman et al., 1996, p. 2). From a dynamic perspective, Rosenberg (1982, p. 120) sees this as 'a learning process' in the generation of new technical knowledge. R&D initially involved purely corporate in-house learning, which all major corporations set up after the Second World War, whether in the form of the ubiquitous laboratory for manufacturing or more diversely as 'new product development' within the marketing department. In the service sector, the locus of learning activities often occurs in groups called 'business development' or 'technology'. Smaller firms also have R&D activities appearing under the titles of 'design' or 'technical support'. All the above require the exchange of information across organizational boundaries *within* the firm, called a 'closed innovation' system. This system entrenches knowledge and its application into the larger corporate entities.

Since the early 1990s, R&D has increasingly evolved into an 'open innovation' system through distributed innovation processes that leverage knowledge from a broad variety of sources *outside* the firm itself, including university academic research, contracting research from 'centres of excellence', joint venture consortiums, acquiring entrepreneurial firms and licensing of innovations. Thus, boundaries for firms conducting R&D have broadened widely under cost pressures and the evolution of the Internet with its supporting web-enabling technologies. This open distributed system has allowed smaller firms to become more involved in the R&D process (on this subject, see Chapter 14, written by B. Laperche). Further, Bowonder et al. (2005, p. 51) note, '[i]ncreasingly firms are acknowledging that it is difficult for them to create and exploit technological innovations on their own'. Companies in high-tech sectors where there is rapid technological change occurring, as in semiconductors, show greater propensity for embracing the open innovation system and encouraging the use of young small creative R&D firms to keep abreast of the frontiers of knowledge (Miotti and Sachwald, 2003).

Research is scientific or technological investigation that has the potential to lead to an idea or concept for innovation. This research is usually conducted by specific experts in two different stages. Basic research is exploratory with no preconceived outcome or direction, and no clearly identified practical applications, but needs to have present or potential interest to the organization conducting the investigation. This research is associated with scientific discovery or, more generally, knowledge building. Applied research has preconceived goals based on business imperatives related to specific products, processes or service delivery. This research is problem-solving and needs to take basic research into practical applications that have indefinable private and/or social returns that relate to strategic positioning of the organization. In an open innovation system, such research involves learning from other technical experts in the variety of sources identified above.

Development explores the specific potential of a product, process or service within an experimental testing environment. This work needs to be conducted at the interface between the technical experts, logistical production managers and marketing departments. Two stages can be identified. First is blueprints development, where a set of designs for specific outcomes are developed from the theoretical research. This is followed by prototype development, which creates test models for technical feasibility (McDaniel, 2002, p. 80). In all stages of development there is need for continual feedback to research in order to improve the theory. This is an iterative process with many failures and dead ends along the way, but is essentially a linear model of R&D innovation. 'In the end, what companies get out of investments in R&D depends on how they manage different stages of the research process itself, and how they get *other* parts of the organization to contribute to the innovation process' (Jolly, 1997, p. 363, original emphasis).

There is much R&D not conducted under the banner of 'R&D'. Particularly, this is the case in two areas. One is 'informal R&D' undertaken on a less organized and ad hoc basis (for example, troubleshooting on the production line). In many countries, especially large ones, informal R&D is regarded as too difficult to survey (Pavitt, 1994), whereas some small countries like Australia survey all size firms for all forms of R&D (Bryant, 1998, p. 59 fn. 4). The other is service-based and some product-based efforts in innovation which use electronic information technology for investigation and testing. Rosenberg (1982, p. 191) identified this problem a long time ago when he said: 'Software development shares many of the problems of any R&D activity.' Freeman (1994) noted the rise of information technology as an area of innovation itself did not come out of identifiable R&D activities. Bowonder et al. (2002) identify the emergence of e-engineering and e-design for innovation as central to R&D but not identified in current R&D metrics. These wider notions of R&D allow the innovation process to be more open-distributed across a larger group of organizations, both public and private, thus moving away from a linear R&D model and towards a more complex interaction across the private–public space.

The vast majority of formal R&D is conducted by a very small band of giant corporations. The USA leads this process, with the National Science Board reporting that as at 2000, 46 per cent of total industrial R&D funds were spent by 167 firms, each employing more than 25,000 employees. At the other end, 32,000 firms with fewer than 500 employees spent only 15 per cent of these same R&D funds (Baumol, 2005, p. 42).[2] The rest of the developed world has this same broad pattern, but not as skewed to larger firms. Despite such massive R&D funds being spent by larger firms, they have historically contributed only a tiny proportion of significant innovation. Thus, after examining the evidence, Baumol (2005, p. 37, original emphasis) concludes 'that perhaps *most* of the revolutionary new ideas of the past two centuries have been, and are likely to continue to be, provided more often by these independent innovators who essentially operate small-business enterprises'.

Very little independent innovation is funded from specified R&D funds. That leaves R&D following relatively routine goals of incremental innovation aimed at improving aspects like reliability, applicability, flexibility, production capacity and user-friendliness. Baumol (2005) associates this incrementalism with adding significantly to economic growth and being of growing importance in the total innovation process. The problem is that these aspects are the beginning of a slippery slide down to planned obsolescence and the marketing imperative of new product development. Such marketing activity can stimulate effective demand, but at the expense of more significant radical innovation that can improve energy and ecological efficiency (Courvisanos, 2005).

What emerges from this discussion of innovation processes are two opposing forces. One is the short-term 'incrementalism' of large firms that moves the economy along a certain patent-protected technological lock-in path in ever-decreasing marginal improvements. The other is the open distributed system of R&D that incorporates smaller young R&D firms with creative radical input into the R&D processes of large firms. Ideally, one force can complement the other, so that radical innovation can cumulatively link to incremental improvements, as suggested by Baumol (2005). However, if the financial power and intellectual property rights of large firms prevail, the opposing forces are resolved in the direction of short-term gains at the expense of both the small creative firms and long-term, more sustainable, innovation outcomes.

6.3 Investment and financing of R&D

R&D expenditure is often referred to as spending, yet conceptually the funds allocated to R&D should be recognized as investment in the future in the same way as investment in plant and equipment. Even failed R&D projects contribute to the corpus of knowledge by identifying what does not work and creates further problems to be solved with further investigations. Basic research is the most creative stage of R&D investment, with the 'more intractable the problem, the more one is curious about it' (Jolly, 1997, p. 375). In this way, R&D spending is a significant part of what economists now call 'intangible investment' because it is an investment into future production, but the knowledge base that such investment creates is not tangible and obvious as plant and equipment (capital) investment (Webster, 1999).

The financing of R&D plays a crucial role in the determination of R&D investment and, in turn, the path that the innovation process takes in a specific economy. In all advanced economies, R&D by medium-to-large firms is internally financed out of profits, making the R&D process endogenous to business and innovation (Griliches, 1995). As the prime example, the USA has the largest national R&D spending in absolute terms,[3] and 70 per cent of this R&D is funded from retained earnings (National Science Foundation, 2004). This internal funding entrenches the power of 'incrementalism' in the innovation process as it reduces risk and provides strong earnings outcomes (Hall, 1992).

The lack of realized profits by young start-up R&D-based firms means that they 'depend greatly on external sources of funding, especially private-sector venture capital markets' (Tassey, 1997, p. 190). The well-documented failings of venture capital markets to support radical innovation undermine the inherent power of creativity within small innovative R&D firms. Tassey (1997, p. 191) identifies three reasons for venture capital markets failing: (i) cyclical nature of the supply of such funds, so that in downturns when innovation is mostly needed, the venture capital firms provide less funds, (ii) shift to less risky later-stage development with shorter expected times to commercialization; (iii) concentration of venture funding in a very few specific industries only, notably pharmaceuticals, biotechnology and information technology. Also, venture capital is concentrated by size of funding, with the minimum size available being much too large for most start-ups (Hall, 2002, p. 48). Further, Hellmann and Puri (2000) show that venture capital funds are more likely to be obtained by firms which are able to bring products to the market significantly fast (they call them 'innovator firms') compared to other 'imitator firms'.[4]

All forms of investment need to be evaluated on the basis of rates of return expected in the future. The future is uncertain, so any calculation of future expected returns is subject to imprecision. For the financing of capital investments on known products or processes, there are standard forecast techniques for calculation of rates of return. R&D is aimed at generating something new that was previously unknown or not present. Given the complexities involved in this process, the outcomes of R&D investment are subject to much greater fundamental uncertainty with no probability distributions available, such that standard forecast techniques are inappropriate (Davidson, 1991).

Fundamental uncertainty is evident when the elements of R&D investment are broken down into their component parts. The various stages of R&D identified above all have different outputs, each one is difficult to evaluate and has diverse possible outcomes, with greatest uncertainty at the early stages of the R&D process. This renders R&D for radical innovation highly problematic (Hall, 2002, pp. 36–7). Thus, evaluation depends on the judgement of experts at the various R&D stages, both inside and outside the firm. However Mansfield et al. (1972), in a classic study of large corporations, identified that these experts tend (when they are planning) to greatly underestimate development costs, while they greatly overestimate the time taken to produce results. Then, Tidd et al. (2005, p. 218) noted that scientists and engineers in basic and applied 'R' are often deliberately overoptimistic in their estimates in order to give the illusion of a high rate of return to conservative accountants, managers and venture capitalists. Such action tends to exacerbate further the difficulties of long-term funding, skewing the funding of R&D towards short-term and more applied 'D' type projects.

The Mansfield study of project selection in large US firms, comparing forecasts to outcomes, found that the probability of picking winners by the technostructures in these firms was only 16 per cent. Jolly (1997), 25 years later, confirms that the Mansfield results still stand even with the advances in modern computer technology. Thus, despite attempts to manage market and technical uncertainty, the technostructure generally fails. This leads large firms to protect any successful innovations in order to maximize returns over as long a period as possible. Such protection can be legal (like secrecy and patents) and illegal (like cartel arrangements), creating monopoly power for that period of protection.[5] This role of the technostructure can be identified as the dominant influence attempting to quell any possible emerging threats (see Shapiro, 2001).

The technostructure needs to resolve the two competing demands on R&D arising from fundamental uncertainty: maximizing the gains from any successful innovation developed in-house, while focusing on radical innovations for a distinctive competitive edge. Tension exists between the two R&D scenarios. Incremental represents minor improvements on existing products or processes that require little organizational change, while radical innovation represents revolutionary departure from current operations with significantly different skills and capabilities.

On the one side there is the recognition that incremental innovation provides 'extra' profits from successfully introduced radical innovations with only marginal R&D input. Financial managers encourage incremental innovation, since it can be calibrated easier with simple 'rules of thumb' for allocating resources, establishing sunset criteria for projects and using sensitivity analysis based on a known range of assumptions while reducing key uncertainties before commitment (Tidd et al., 2005, pp. 218–20). This is an organizational constraint where financial shackles stymie innovative edge as financial controllers seek short-term gains. Chiesa (2001) in his guide to R&D management argues for commitment of funds for R&D in the incremental innovations of existing corporate technologies. New technological opportunities, Chiesa explains, require difficult and risky overall corporate strategic planning changes that need first to overcome organizational constraints by realigning a firm's strategic plan to some core technological focus. For example, are automobile manufacturers in the business of individual people movers or the petrol-driven vehicles? The latter is the incremental approach that limits strategic focus to a narrow technological focus.

Incrementalism is further entrenched by marketing efforts and monopoly power. Professional R&D executives recognize the role of marketing in its interaction with lead users of the products in setting the R&D agendas. This is done not only by standard marketing 'research' surveys, but increasingly more prevalent has become collaboration with lead users on finding what such users 'need' to improve use of their products, for example mountain bikes and computer software (von Hippel, 2005). Monopoly power of secrecy and property rights aim to stifle radical innovation being conducted by smaller entrepreneurial firms or even in-house radical ideas which threaten the current strong market position of the dominant firm(s). Lessig (2004) has argued this case very persuasively with strong empirical evidence from the media industry. For example, RCA squashed all attempts by their R&D engineer, Edwin Armstrong, to introduce the higher-quality FM radio band; all RCA wanted was to protect their monopoly of the

AM band by reducing the static noise on the AM band (Lessig, 2004, pp. 3–7).

On the other side, there is evidence that radical innovations are significantly more likely to be commercially successful (Ettlie and Rubenstein, 1987). This is because the accumulated firm-specific intangible knowledge for future opportunities (first mover advantage) tends to be greater the more radical the innovation. The difficulty is in assessing which ideas will eventually succeed and having to pursue many on the expectation that one will succeed. There is no probability distribution and thus no calculable risk assessment that can be made for successful radical innovations. This biases the financing of R&D away from such radically creative ideas. Ettlie (2000, p. 40) estimates that only 6–10 per cent of all new successful products are radical, while successful radical processes are even scarcer. Also, radical processes tend to follow radical products with a lag, but then both technologies become embedded and lose their 'cutting edge' (Abernathy and Utterback, 1978). Focusing on radical innovations not only requires a considerable shift in skill capability and organizational structure, but it also introduces the threat of new entrants (some very large with 'deep pockets') into the industry who are prepared to diffuse the innovation. Concentration on radical R&D requires brave foresight and supportive long-term financing on the part of established business.

Up to the early 1960s R&D was funded directly from central corporate sources. Since then, a growing movement has emerged to fund R&D from contracts between the R&D division and other internal and external business 'groups'. For example, Philips began in 1990 to have its central head office-based funding across its five laboratories around the world reduced to one-third, with the remainder coming from contracts from business groups (Jolly, 1997, p. 346). This distributed funding trend threatens creative R&D in radical innovations and tends to support incremental innovation driven by contract-based strategic marketing needs. Philips, realizing this, modified their funding structure in 1994, requiring roughly half of the two-thirds controlled by business groups to be 'devoted to immediate product development; the remaining half has to be for longer-term capability development in certain technology clusters, such as signal processing for TVs. Typically, this part is funded by more than one business group as well' (Jolly, 1997, p. 349). This is the nub of the funding R&D dilemma, determining how R&D strategies address both short-term market-based needs and long-term knowledge-accumulation needs. This all depends on the valuation of strategic intent by the firm undertaking R&D.

The valuation of strategic intent is influenced by two factors. One is the life cycle of the current radical innovation. During the early stages of a successful innovation, incremental changes out of R&D result in substantial gains for the firm and in terms of social benefit as the innovation is adapted and diffused. Then as the innovation matures, R&D tends to suffer diminishing returns in terms of new knowledge and new applications. At this mature stage defensive R&D efforts aim to maintain market position (Bar, 2006). The other factor is the size of the firm. Ettlie and Rubenstein (1987) examined 348 US manufacturing firms and identified smaller firms (up to 1000 employees) as introducing at the same rate both radical and incremental new products, then as firms increased their size up to 11,000 employees their greater size tended to promote more radicalness. When firms became very large (greater than 11,000 employees) and the technostructure became powerful, there was a clear lack of radical product innovation despite often very large R&D units.

The pressure towards short-term R&D financing within medium-to-large corporations and the enormous financial difficulties smaller firms have in being able to enter the high-cost, high-risk R&D world, are the basic financing features of private-sector R&D. Given the high economic and social rate of return that R&D bestows upon a nation,[6] it is not surprising that governments of all political colours have entered into the R&D sphere with specific public policies.

6.4 R&D public policies

Central to all nations' industrial and innovation public policy strategy is the approach governments take to funding and supporting R&D. Gerschenkron (1962) associates such strategy with the 'late' industrial development stage of the global economy. The economics literature has identified four rationales for such emphasis on supporting what is essentially a private sector activity.

The first is the neoclassical supply-oriented competition concerns arising from 'market failure', developed most notably by Arrow (1962). This argument is based on inadequate return for the private sector in R&D due to few and uncertain pay-offs from basic research. As noted, large-firm R&D tends to support short-term incrementalism, because there is limited market-based encouragement for more uncertain new technologies in less powerful industries and firms where the scale of R&D is too low to generate the critical mass of new knowledge. Also, duplication by competitors tends to quickly undermine any competitive edge established by the initiator (free rider issue). From the nation's standpoint, these problems of market failure lead to underinvestment in R&D. Tassey (1997, p. 88) identifies three types of market failures that imply different public policy responses. Type I, general risk aversion and lack of small business involvement, leads to tax incentives and subsidies that attempt to reduce the general costs of R&D. Type II, R&D-specific technical risk, leads to direct funding of basic (or generic) research through public-based research institutions, assignment of property rights and funding consortia. Type III, market-related risk, leads to providing a strong infrastructure base to R&D, with effective technical systems, artefacts and standards which allows R&D to prosper. From the firm's perspective, market failure rationale lowers the risk and subsidizes the costs of R&D.

The second rationale centres on national security issues as developed by Gansler (1980). Ability to be self-sufficient in circumstances of secrecy on defence (and space programme) strategies drives this concern (offensive-based). It is bolstered by concerns of being cut off or refusal to trade during military conflicts (defensive-based). R&D spending, due to secrecy and lack of direct civilian applicability, cannot be supported in private markets. This R&D is financed by the public sector, but developed in the private sector, with the use of procurements to drive down R&D costs. In the long run the knowledge gained provides a platform for new civilian capabilities far into the future (for example computers, GPS, commercial space travel). This has been the case throughout history, but clearly at different rates of civilian uptake (White, 2005).

The third rationale is based on evolutionary economics, centred on a systems approach that rejects the linear model of R&D. The national innovation system is a set of institutions whose complex interaction via clusters, collaborations and networks across the public–private sector space determines the extent of innovative performance (Nelson, 1993). In this system, R&D forms the foundation of knowledge and its applicability for innovation. However, systemic failures in private sector R&D due to incrementalism, lock-in, transitional problems, poor knowledge-based infrastructure, and inappropriate conventions and institutions justify the need for national governments to overcome these failures in a strategic way (Smith, 1998). For example, private-sector R&D support for small firms with single innovation ideas are hard to justify on financial grounds because the chances of success before any patents expire are very low (Legge and Hindle, 2004, p. 337).

The final rationale is based on environmental concerns. Exhaustion of non-renewable resources, pollution and greenhouse emissions threaten the environment's ecosystem viability, while markets do not reflect the ecological value of sustainability of human and other life on this planet. Thus, there is a need for public finance and support of R&D on decentralized alternative low-emission energy sources and reducing pollution (McDaniel, 2002, p. 85). Neoclassical and evolutionary economists could claim this argument for their respective market or systemic failure arguments; however, ecological economists see the ecosystem overriding both such approaches. A market failure approach can merely encourage the public support of R&D into costly and unsustainable 'end-of-pipe' technological solutions. A systemic failure approach to work from this environmental perspective needs R&D that has clear ecological directions and rules that allow for adaptation and incremental change towards a decentralized sustainable ecosystem (rather than support, for example, of massive centralized nuclear power and corporate genetic engineering; see Skea, 1994).

Two types of R&D public policies are possible, passive and active (Legge and Hindle, 2004, pp. 237–50). Passive policies respect laissez-faire market solutions by attempting to override market failures, giving markets a better chance to work effectively. This would involve intellectual property rights (IPRs) protection of R&D innovation to overcome the free rider issue, and providing broad R&D rebates, subsidies and incentives in order to reduce risk and support scale economies. This is the neoclassical approach to R&D public policies. Cannon (2005) explains that the USA, as the leader in R&D, has strong preference for passive R&D policies, and notes that the four successful R&D instruments are (in order of importance): tax relief, defence support, patent protection and college education. The paradox of passivity by not picking winners and yet supporting massive defence R&D does not seem to be apparent in Cannon's analysis, but this is to be expected from the perspective of the dominant structures that inhabit R&D in the USA. A more recent variation of this neoclassical 'passive' approach has been policies to shift R&D support from large corporations towards small business (through programmes for technology start-up companies like pre-seed funding and incubators). Though the conservatives could suggest this change is due to market failure as large corporations override the market, such a post hoc rationale undermines the whole passive approach and leads directly to active policies which are antithetical to the conservatives.

Active policies aim to directly intervene in order to influence the direction and extent of R&D innovation. Sectoral R&D assistance to specific industries aims to address concerns of the lack of innovation in this area (for example, CSIRO as the Australian public research body in support essentially of the agricultural sector). Selective public investment in research infrastructure (for example synchrotrons, technology parks, cooperative university–business research centres), subsidies in specific areas of concern (environment, social groups, non-urban regions) and public sector procurement of R&D (as in the defence industry) all provide direction as part of public policy support. All rationales bar the neoclassical tend to support such active policies, with the particular direction of R&D support up for debate at the political level (centralized authority or democratic grass roots). The proponents of such active policies argue on the basis that these are emergent influences which they would want to see eventually dominate.

In reality, R&D public policies end up being a mix of both passive and active, depending on the political trajectory that a nation has traversed over the previous 50 years. The trend of R&D policies reflects the rationale which is being championed by the political powers at the time. There are, however, some theoretical limitations to R&D support by the state. In relation to subsidy/incentive-type support, successful R&D innovations end up benefiting the private sector firm involved twice, once from financial support and second from profits of the innovation often with state-endorsed monopoly control through IPRs. Concern also exists as producers of R&D get exclusive benefits of the IPRs, when often it is users who generate the innovative ideas even though all benefits and IPR protection go to the generally large patent-holding producers (see von Hippel, 2005).[7]

Questions are also raised about governments' attempts to 'boost' R&D when it is used merely as a marketing tool for incremental innovation (How many blades can you place on a shaver?). This is supported by evidence that incumbent enterprises, with minor innovative activity, benefit most from such R&D public support during long economic expansions, whereas new firm start-ups are triggered by economic contraction and unemployment supported by university research in particular (Audretsch and Acs, 1994). At the other extreme, government support of radical innovation is also problematic. There is the growing neo-liberal influence in many Western economies to encourage support for small-based entrepreneurial start-ups based on some 'exaggerated claims of their role in innovation' (Legge and Hindle, 2004, p. 247), when in fact the vast majority of entrepreneurial start-ups are extensions of work conducted prior to start-up (Bhidé, 2000). Further, Åsterbo (2003) shows evidence of unrealistic optimism in a sample of 1091 independent inventions, with only between 7 and 9 per cent reaching the market and 60 per cent of them obtaining negative returns.[8]

Empirical evidence on R&D support is mixed. Bloom et al. (2002) draw the conclusion from a nine major OECD-country study that generally R&D tax credits have had a significant effect. However, other studies have found several problems with this form of incentive; in particular, it applies to only new R&D, criteria are stringent, there is no distinction between R&D spending and success rates, productivity effects are varied, and it ignores the increasingly important role of collaborations (Ettlie, 2000, pp. 298–300). Active policies like selective investment in incubators and technology parks have had varying success, depending on how well targeted the policy is, how well it is administered and monitored; then there is the level of synergy of companies involved with similar and complementary endowments. Finally there is the motivation of the participants themselves in these research infrastructures. Australia has been notable for selective funding of investment in two major successful innovation-based research infrastructures: CSIRO and AIS (Australian Institute of Sport). Both have been models that have been studied and copied around the world; however, Australia's natural and cultural endowment in agriculture and sport have much to do with this success (Fox, 2001).

A third of OECD countries (all small economies) have the public sector as their major source of R&D funding, also all less developed economies depend on government for R&D. From this it can be noted that higher education and government sectors perform almost 30 per cent of all R&D (OECD, 2005). However, the track record for active R&D policies is mixed. In concert with major security concerns arising from the collapse of the World Trade Center on 11 September 2001, the military rationale ensures continued success on the innovation front. The other two active-based rationales have had mixed results, with proponents pointing to countries like the Netherlands who have been able to develop innovation policies for sustainable development. Such success has only been achieved on the back of a major environmental crisis that the population as a whole has come to accept (Courvisanos, 2005).

6.5 R&D and the global financial system

An important task for political economy is to appreciate the implications of the new globalizing financial order. Argitis and Pitelis (2006) articulate the slow integration, since the collapse of the Bretton Woods Agreement in 1971, of the advanced economies into 'largely unregulated financial markets'. These markets are characterized by no capital controls and flexible exchange rates, resulting in mobile financial capital that seeks to obtain the highest returns available on a global scale. The resulting economic and financial instability arising from such a global system has been well documented.[9] This new financial order has significant consequences for R&D, the innovation process and economic growth.

Stiglitz (1999) identified that the increased pattern of instability in the increasingly liberalized global financial system implies greater riskiness for investment, and in particular R&D investment. This implies three major R&D consequences that arise from the analysis already discussed. First, this exacerbates the tendency by large firms to fund short-term R&D projects. Second, during economic downturns R&D is curtailed even though it is a crucial time when such investment can stimulate growth out of a trough purely by creating effective demand (whether the R&D is successful is immaterial, it is the spending on R&D that stimulates demand). Third, the ability of young R&D companies to access external funding with 'patient money' is significantly reduced.

The short-term perspective employed under global financial liberalization has a further implication of reducing the capacity of the innovation process to gain the most out of R&D spending. Investing in long-term R&D projects has the cumulative benefit of adding significant new demand created from technological developments out of successful commercialization of R&D. However, concentrating investment in short-term R&D limits the ability to cumulatively add to demand, allowing only for incremental improvement in replacement of current products and processes rather than adding another technical progress layer to the economy. This short-term 'incrementalism' limits economic growth. Further, this perspective also works against the development of long-term R&D projects into emissions-reducing radical innovations for sustainable economic development.

The other major consequence of global financial liberalization that financial instability causes is the increasing shift of capital funds to 'rentier-led' speculative short-term portfolio investments (Argitis and Pitelis, 2006, p. 77) (a subject dealt with by J. Galbraith in Chapter 1 and J. Sawyer in Chapter 4). This shift is at the expense of long-term productive investment in R&D and related capital investment. Rent capture in an unstable financial environment displaces value creation from technological progress, limiting the role of R&D to gaining and exploiting the control of intangible knowledge. Archibugi and Michie (AM) (1995) develop a taxonomy to account for these forces of globalization. Still at the centre of R&D activity are the large firms with their domestic base that enables 'global generation of technology' that leads to new IPRs. OECD (2005) identifies that only 12 per cent of the world's large firms conduct R&D outside their home country, compared to around 25 per cent equivalent share of production. Empirical studies by Coe and Helpman (1995) and Verspagen (1997) show that large multinational enterprises transfer this technological knowledge as significant international spillovers through foreign direct investment (FDI) to the second level. This second level, AM identify as 'global technological collaboration' based on transferring technology through strategic alliances to further develop and apply technological advantages. This allows R&D from the advanced economies to be absorbed through what Sachs (2003) calls the 'technological diffuser' group of countries. This group consists of the newly technologically emerging regions limited to eastern and central China, large parts of India, Latin America and some parts of Eastern Europe.

The third level of the AM taxonomy is 'global exploitation of technology'. This is where rent capture is clearly observed as the exploitation of corporations' technological advantage across all markets that have the wherewithal to purchase the new technology. Sachs (2003) notes that this globalization process excludes about one and a half billion people who are marginalized completely from technological advance, specified as the Andean region of South America, almost all sub-Saharan Africa and large parts of Central and South Asia.

The above globalization forces have resulted in a situation that on average foreign-based production is less innovation-intensive than home production, with firms from smaller countries generally having higher shares of foreign innovative activities. Most R&D performed outside home sites occurs in the USA and Germany, with a growing trend in biotechnology and IT for European firms to conduct R&D in the USA so as to access local skills and knowledge (Tidd et al., 2005, pp. 211–13). Meyer-Krahmer and Reger (1999) characterize R&D as being in the dominant 'Triadization' structure, involving companies from the USA, the European Union and Japan. OECD (2005) figures indicate at the broader level of 30 leading economies (excluding China and India), that well over 16 per cent of total R&D expenditure is performed abroad by foreign affiliates. The picture that emerges is a complex mosaic of rising internationalization of R&D but with limited 'techno-globalism'.[10]

The Basel Accord capital adequacy regulatory framework aims to protect the liberalized global financial system from the worst excesses of speculation.[11] In June 2004, the Basel II Capital Accord was agreed to by the developed economies' financial regulators and is being progressively introduced through the world's major nations. Basel II was devised to ensure that the capital adequacy measures are much more risk sensitive by itemizing and quantifying several more categories of risk in an internal ratings-based system of lending. In theory, this should encourage financial institutions to be less cautionary and finance R&D activities on the basis of qualitative criteria incorporating intangible knowledge assets. However, as Elisa Ughetto discovers in her study of Italian banks (see Chapter 10 of this volume), the restrictive and conservative regulations of the International Accounting Standards adopted by firms ignore intangible investment, thus limiting significantly the innovation value of this new accord in practice.

6.6 Conclusion

Analysis of the academic R&D literature above reveals a critical political economic imperative that is a theme of this volume; tension between innovation and predation. Capitalists have the difficult task of determining whether R&D should finance long-term significant issues (in particular, energy and ecological concerns), or instead concentrate finance on short-term returns out of routine-based incremental innovation. The greater fundamental uncertainty of the former is exacerbated in a global financial system which is increasingly being liberalized and more unstable. Ability to capture strong rents on the back of short-term incremental R&D militates against support for radical innovation that is inherently more uncertain. Thus, the predation element dominates in the form of technostructures that finance short-term R&D projects and venture capitalists that are not able to adequately finance young start-up R&D firms. Continuance of this convention towards short-term R&D financing will threaten the high social returns to R&D investment that have been identified through the post-Second World War period. Attempts by regulators at both national and international levels to encourage a longer-term perspective are fighting against the strong market forces that encourage such short-termism.

The story that emerges from this analysis is complex, such that the above forces for rent capture can only work for a certain amount of time. As marginal returns in incremental innovation continue to decline, the pressures for significant radical innovation rise. This pressure is, however, contained within a tightly controlled system where R&D public sector support tends towards incumbent enterprises. Effort to support start-ups is difficult, given that the vast majority of them will fail. For government authorities, supporting large firms with an established R&D

record is much more cost-effective and delivers strong political kudos to the political party in power. The approach that the Basel II Accord encourages for the development of more radical and innovative R&D is a significant 'straw in the wind'. This approach can be combined with the strong ecological technology-based R&D that is emerging out of climate change-induced policies in Europe and parts of the USA. Together they have the potential to shift the R&D process to long-term private returns with a consequent important rise in the social returns of R&D. For this approach to work there is a need to first clearly appreciate the political economy forces, outlined in this chapter, that have been amassed in favour of short-termism. Then, to introduce policies that will discourage the conservative forces of incrementalism, while encouraging the forces that can bring significant long-term benefits to the broad global community, rather than to the narrow powerful global financial community.

Notes

1 I appreciate the generous support of the Centre for Strategic Economic Studies at Victoria University for Visiting Fellow status during the writing of this chapter. Thanks go to Phil O'Hara (Curtin University), Wifred Dolfsma (Erasmus University) and Paul Ramskogler (Vienna University of Economics) for their comments on earlier drafts.
2 This ignores the entire informal R&D conducted by all firms as described above which, if included, would greatly boost the total R&D activity without probably altering the proportions.
3 R&D spending in the USA (in US dollar terms) is greater by a factor of two ahead of second-placed Japan, and a factor of five ahead of third-placed Germany (Meginnson, 2001, p. 37).
4 This seems to be a peculiar nomenclature, possibly based on the view that bringing the product quickly to the market is the innovative criterion. Most *really* innovative products (and processes) actual take a reasonably long time to be commercialized (see Jolly, 1997).
5 For example, in June 2004 the US Congress extended copyright on music from 50 to 95 years in order to protect the commercial interests of recorded popular 'rock' music (the Sonny Bono amendment). This date was (*not* coincidentally) one month before the first major 'rock' recording, Elvis Presley's 'That's All Right Mama', was about to run out of copyright since it was released on 19 July 1954 (*The Sunday Age*, 2004).
6 Griliches (1992) shows empirically that the social rate of return on investment in R&D is greater than the private rate, and this is supported by more recent work by Lach (2005) on Israel – one of the top five R&D spending nations in terms of percentage GDP.

7   The problem of IPRs 'being hijacked by larger firms, particularly for strategic purposes' is a public policy concern addressed by Dolfsma (2006, p. 339).
8   Audretsch (1995, p. 122) places the above discussion into the US context by stating that 'although divergences in beliefs regarding (potential) innovations may induce a greater amount of start-up activity, the likelihood of any new firm actually surviving and having a substantial impact on the industry is relatively low. Thus we observe a relatively high number of entrepreneurial or young firms, whose impact is, on average, relatively negligible.'
9   Arestis et al. (2003), Strange (1998), Gill (1995), Helleiner (1994) and Frieden (1991) all describe this financial globalization and its instability from differing but critical perspectives.
10 See also OECD (1998) for details of these internationalization aspects of R&D.
11 The Bank of International Settlements (BIS), based in the Swiss town of Basel, fosters cooperation among the central banks and regulatory agencies in the global financial system in pursuit of monetary and financial stability. For an introduction to the Basel Accord, see Wikipedia website: http://en.wikipedia.org/wiki/Basel_Accord. For more detailed research papers on BIS and the Basel Accord, see BIS website: http://www.bis.org/forum/research.htm.

References

Abernathy, W. and J. Utterback, 'Patterns of Industrial Innovation', *Technology Review*, 80 (1978), 40–7.
Archibugi, D. and J. Michie, 'The Globalisation of Technology: a New Taxonomy', *Cambridge Journal of Economics*, 19 (1995), 121–40.
Arestis, P., M. Nissnke and H. Stein, 'Finance and Development: Institutional and Policy Alternatives to Financial Liberalisation', The Levy Economic Institute, Working Paper 377 (2003). Accessed at http://www.levy.org.
Argitis, G. and C. Pitelis, 'Global Finance, Income Distribution and Capital Accumulation', *Contributions to Political Economy*, 25 (2006), 63–81.
Arrow, K. 'Economic Welfare and the Allocation of Resources for Invention', in R. Nelson (ed.), *The Rate and Direction of Inventive Activity: Economic and Social Factors* (Princeton, NJ: Princeton University Press, 1962), pp. 602–25.
Åsterbo, T., 'The Return to Independent Invention: Evidence of Unrealistic Optimism, Risk Seeking or Skewness Loving?', *The Economic Journal*, 113 (2003), 226–39.
Audretsch, D., *Innovation and Industry Evolution* (Cambridge, Mass.: MIT Press, 1995).
Audretsch, D. and Z. Acs, 'Entrepreneurial Activity, Innovation and Macroeconomic Fluctuations', in Y. Shionaya and M. Perlman (eds), *Innovation in Technology, Industries and Institutions: Studies in Schumpeterian Perspectives* (Ann Arbor: University of Michigan Press, 1994), pp. 173–83.
Bar, T., 'Defensive Publications in an R&D Race', *Journal of Economics & Management Strategy*, 15 (2006), 229–54.
Baumol, W., 'Education for Innovation: Entrepreneurial Breakthroughs versus Corporate Incremental Improvements', in A. Jaffe, J. Lerner and S. Stern (eds),

Innovation Policy and the Economy, vol. 5 (Cambridge, Mass.: MIT Press and National Bureau of Economic Research, 2005), pp. 33–56.

Bhidé, A., *The Origin and Evolution of New Businesses* (Oxford: Oxford University Press, 2000).

Bloom, N., R. Griffith and J. van Reenen, 'Do R&D Tax Credits Work? Evidence from a Panel of Countries 1979–97', *Journal of Public Economics*, 85 (2002), 1–31.

Bowonder, B., P. Sudhakar and D. Wood, 'E-Engineering: Redefining the Boundaries of the Firm', *International Journal of Information Technology and Management*, 1 (2002), 32–49.

Bowonder, B., J. Racherla, N. Mastakar and S. Krishnan, 'R&D Spending Patterns of Global Firms', *Research-Technology Management*, 48 (2005), 51–9.

Bryant, K., 'Evolutionary Innovation Systems: Their Origins and Emergence as a New Economic Paradigm', in K. Bryant and W. Alison (eds), *A New Economic Paradigm? Innovation-based Evolutionary Systems* (Canberra: Department of Industry, Science and Resources, 1998), pp. 53–84.

Burgelman, R., M. Maidique and S. Wheelwright (eds), *Strategic Management of Technology and Innovation*, 2nd edn (New York: McGraw-Hill, 1996).

Cannon, P., 'Why We Do R&D (a Practitioner's Tale)', *Research-Technology Management*, 48 (5) (2005), 10–11.

Chiesa, V., *R&D Strategy and Organisation: Managing Technical Change in Dynamic Contexts* (London: Imperial College Press, 2001).

Coe, D. and E. Helpman, 'International R&D Spillovers', *European Economic Review*, 39 (1995), 859–87.

Courvisanos, J., 'A Post-Keynesian Innovation Policy for Sustainable Development', *International Journal of Environment, Workplace and Employment*, 1 (2005), 187–202.

Davidson, P., 'Is Probability Theory Relevant for Uncertainty? A Post Keynesian Perspective', *Journal of Economic Perspectives*, 5 (1991), 129–43.

Dolfsma, W., 'IPRs, Technological Development, and Economic Development', *Journal of Economic Issues*, XL (2006), 33–42.

Ettlie, J., *Managing Technological Innovation* (New York: John Wiley & Sons, 2000).

Ettlie, J. and A. Rubenstein, 'Firm Size and Product Innovation', *Journal of Product Innovation Management*, 4 (1987), 89–108.

Fox, J., 'Why is it so Difficult to Develop Great Ideas and Inventions in Australia', in *The Alfred Deakin Lectures: Ideas for the Future of a Civil Society* (Sydney: ABC Books, 2001), pp. 228–38.

Freeman, C., 'The Economics of Technical Change: a Critical Review Article', *Cambridge Journal of Economics*, 18 (1994), 463–514.

Frieden, J., 'Invested Interests: the Politics of National Economic Policies in a World of Global Finance', *International Organization*, 45 (1991), 425–51.

Galbraith, J., *The New Industrial State* (Boston: Houghton Mifflin, 1967).

Gansler, J., *The Defence Industry* (Cambridge, Mass.: MIT Press, 1980).

Gerschenkron, A., *Economic Backwardness in Historical Perspective: a Book of Essays* (Cambridge, Mass.: Belknap Press of Harvard University Press, 1962).

Gill, S., 'Globalization, Market Civilization and Disciplinary Neoliberalism', *Millennium: Journal of International Studies*, 24 (1995), 399–423.

Griliches, Z., 'The Search for R&D Spillovers', *Scandinavian Journal of Economics*, 94 (1992), S29–47.

Griliches, Z., 'R&D and Productivity: Econometric Results and Measurement Issues', in P. Stoneman (ed.), *Handbook on the Economics of Innovation and Technological Change* (Oxford: Blackwell, 1995), pp. 53–89.

Hall, B., 'Investment and Research and Development at the Firm Level: Does the Source of Financing Matter?' National Bureau of Economic Research (NBER), Working Paper 4096 (1992). Accessed at http://papers.nber.org/papers.html

Hall, B., 'The Financing of Research and Development', *Oxford Review of Economic Policy*, 18 (2002), 35–51.

Helleiner, E., *States and the Re-emergence of Global Finance* (Ithaca, NY: Cornell University Press, 1994).

Hellmann, T. and M. Puri, 'The Interaction between Product Market and Financing Strategy: the Role of Venture Capital', *Review of Financial Studies*, 13 (2000), 959–84.

Jolly, V., *Commercializing New Technologies: Getting from Mind to Market* (Boston: Harvard Business School Press, 1997).

Lach, S., 'Do R&D Subsidies Stimulate or Displace Private R&D? Evidence from Israel', *Journal of Industrial Economics*, 50 (2005), 369–90.

Legge, J., and K. Hindle, *Entrepreneurship: Context, Vision and Planning* (Basingstoke: Palgrave Macmillan, 2004).

Lessig, L., *Free Culture: How Big Media Uses Technology and the Law to Lock Down Culture and Control Creativity* (New York: Penguin, 2004).

McDaniel, B., *Entrepreneurship and Innovation: an Economic Approach* (Armonk, NY: M.E. Sharpe, 2002).

Mansfield, E., J. Raporport, J. Schnee, S. Wagner and M. Hamburger, *Research and Innovation in the Modern Corporation* (London: Macmillan, 1972).

Meginnson, W., 'Towards a Global Model of Venture Capital?', University of Oklahoma, 31 December draft (mimeo, 2001). Accessed at http://facultystaff.ou,edu/M/William.L.Megginson-1

Meyer-Krahmer, F. and G. Reger, 'New Perspectives on the Innovation Strategies of Multinational Enterprises: Lessons for Technology Policy in Europe', *Research Policy*, 28 (1999), 751–76.

Miotti, L. and F. Sachwald, 'Co-operative R&D: Why and with Whom? An Integrated Framework of Analysis', *Research Policy*, 32 (2003), 1481–500.

National Science Foundation, *Product versus Process Applied Research and Development by Selected Industry* (Arlington, Va: 2004). Accessed at: http://www.nsf.gov/sbe/srs/iris

Nelson, R. (ed.), *National Innovation Systems: a Comparative Study* (Oxford: OUP, 1993).

OECD, *Internationalisation of Industrial R&D: Patterns and Trends* (Paris: Organization of Economic Cooperation and Development, 1998).

OECD, *OECD Science, Technology and Industry Scoreboard 2005* (Paris: Organization of Economic Cooperation and Development, 2005).

Pavitt, K., 'Key Characteristics of Large Innovating Firms', in M. Dodgson and R. Rothwell (eds), *The Handbook of Industrial Innovation* (Cheltenham: Edward Elgar, 1994), pp. 357–66.

Rosenberg, N., *Inside the Black Box: Technology and Economics* (Cambridge: Cambridge University Press, 1982).

Sachs, J., 'The Global Innovation Divide', in A. Jaffe, J. Lerner and S. Stern (eds), *Innovation Policy and the Economy*, vol. 3 (Cambridge, Mass.: MIT Press and National Bureau of Economic Research, 2003), pp. 131–41.

Schumpeter, J., *The Theory of Economic Development* (Cambridge, Mass.: Harvard University Press, 1934) [German original 1911].

Schumpeter, J., *Capitalism, Socialism and Democracy* (New York: Harper & Row, 1942).

Shapiro, C., 'Navigating the Patent Thicket: Cross Licenses, Patent Pools and Standard Setting', in A. Jaffe, J. Lerner and S. Stern (eds), *Innovation Policy and the Economy*, vol. 1 (Cambridge, Mass.: MIT Press and National Bureau of Economic Research, 2001), pp. 119–50.

Skea, J., 'Environmental Issues and Innovation', in M. Dodgson and R. Rothwell (eds), *The Handbook of Industrial Innovation* (Cheltenham: Edward Elgar, 1994), pp. 421–31.

Smith, K., 'Innovation as a Systemic Phenomenon: Rethinking the Role of Policy', in K. Bryant and W. Alison (eds), *A New Economic Paradigm? Innovation-based Evolutionary Systems* (Canberra: Department of Industry, Science and Resources, 1998), pp. 17–51.

Stiglitz, J., 'Reforming the Global Economic Architecture: Lessons from Recent Crises', *The Journal of Finance*, 54 (1999), 1508–21.

Strange, S., *Mad Money* (Manchester: Manchester University Press, 1998).

Tassey, G., *The Economics of R&D Policy* (Westport: Quorum Books, 1997).

The Sunday Age, 'Copyright Rock', 25 July 2004, Agenda, p. 6.

Tidd, J., J. Bessant and K. Pavitt, *Managing Innovation*, 3rd edn (Chichester: John Wiley & Sons, 2005).

Verspagen, B., 'Estimating International Technology Spillovers Using Technology Flow Matrices', *Weltwirtschaftliches Archiv*, 133 (1997), 226–48.

Von Hippel, E., *Democratizing Innovation* (Cambridge, Mass.: MIT Press, 2005).

Webster, E., *The Economics of Intangible Investment* (Cheltenham: Edward Elgar, 1999).

White, M., *Fruits of War: How Military Conflict Accelerates Technology* (London: Simon & Schuster, 2005).

7
Finance and Intellectual Property Rights as the Two Pillars of Capitalism Changes

George Liodakis

7.1 Introduction

It is hard to overemphasize the significance of technology and innovation in the context of the currently evolving technological revolution and the concurrent fundamental restructuring, ushering in the newly emergent stage of *transnational* or *totalitarian capitalism* (see Liodakis, 2005, 2006). In the same way it is also wrong to consider that this revolution of new technologies and the corresponding social restructuring of capitalism is an entirely new and historically unprecedented phenomenon (see Noble, 1977, pp. xxiii–xxiv).

The fundamental global restructuring of capitalism under way during the last three decades, constitutive of the newly emergent stage of capitalism, is itself the outcome and response to the deep structural crisis of capitalist overaccumulation, which came to the forefront during the early 1970s (see Perelman, 2003; Liodakis, 2005). The reproduction and survival of capitalism, at its current stage, requires drastic counteractive measures be taken for the tendency of the rate of profit to fall, and in this respect technological innovation and development acquire a crucial importance. But in order to properly assess the role of technology in the restructuring of capitalism and as a means to overcome the recent deep crisis, we need to transcend the prevalent conception of technological determinism (see Noble, 1977, p. xxvi; 1986, pp. x–xi). It is in other words necessary, contrary to this conception of technological determinism, to penetrate beneath the apparent phenomena of the current social restructuring of capitalism and the associated technological transformations, in order to conceive more fully the real essence of technology itself (see Noble, 1977, p. xxii; Liodakis, 2003), to more adequately grasp the determinants of technology, its real trends and their implications, and to understand that technology can be neither exogenous to the capitalist accumulation process, nor socially neutral (see Noble, 1995; Liodakis, 2006, and relevant references). Speaking of technological developments, it is methodologically important to stress, more generally, that these developments should be understood in a close dialectical association with the social restructuring of society. This conception can be safely grounded in the dialectic interaction between the forces and the social relations of production stressed by Marx and other Marxist researchers (see Noble, 1977, p. xix, among others).

It is also worth recalling that the dynamics of capitalist accumulation and development convinced Marx about the ever-increasing role and significance of scientific and technical knowledge. However, as Marx (1973, p. 706; 1967, III, pp. 439–40) stressed, the rising productivity of capitalist production misleadingly appears, not as the product of the collective knowledge and cooperation of workers, but rather as a product of capital. He also stressed that, along with the increasing significance of science and technical knowledge, capital shows a tendency to dissociate labour from the productive process and reintegrate the two on different terms, thus allowing a maximum surplus value extraction. This led Marx to point out that 'modern industry ... makes science a productive force distinct from labour and presses it into the service of capital' (as cited in Noble, 1977, p. xxiv). It is also crucially important to point out the twofold significance of the development of the productive forces, as they are both reinforcing and undermining the existing social relations of production. The growing socialization of production, the rapid development of science and technology, and the increasing undermining of capitalist property relations do indeed seem, nowadays, to create the preconditions for important institutional changes (see also Liodakis, 2003; Carchedi, 2005).

It should be noted further that the dissociation of labour from knowledge and productive intelligence, mentioned above, also constitutes a core element of the Taylorist movement in the early part of the twentieth century (see Braverman, 1974; Perelman, 1998, pp. 41–3), as well as the recent fragmentation and manipulation of genetic resources by modern biotechnology (see Liodakis, 2003; Carchedi, 2005), and the present-day trend for the Taylorization of scientific research and education (Noble, 1998). On these grounds, a huge surge has taken place during recent decades in the expansion and protection of intellectual property rights (IPRs). As this institutional change plays a crucial role in the technological development and social restructuring of capitalism, we need to analyse this factor more specifically. Contrary to authors who overstress some observable evolutions in the institutions of private property (see Rifkin, 2000), we have elsewhere argued that the significance of private property is largely maintained in the new stage of capitalist development (Liodakis, 2006), and this applies not only to material property, but also to knowledge and intellectual property (see also UNESCO, 2005, p. 116).

Apart from the important role of private property and IPRs, finance has also emerged as one of the major determinants of both technological development and social restructuring. It is necessary, therefore, to investigate more specifically the role of finance and its implications for technological innovation and development. In this respect, it is important to stress that, although '[a]ttention to the historically varying forms taken by competition and the operation of the law of value is crucial' (Kincaid, 2005, p. 113), 'the financial markets have become increasingly central in processes of capital allocation' (ibid.) over the past 30 years, and the role of the financial and credit system is apparently crucial for the inducement and development of technological innovations. However, what is more significant in understanding current development is the core process of surplus value extraction and appropriation by capital rather than its apparent market forms. As argued, although '[c]apitalism as a mode of production has gone through a number of fundamental structural mutations ... in explaining capitalism historically, what is essential in Marxist political economy is not value-form theory, but Marx's dynamic theorization of capitalist competition, summarized in the shorthand term *law of value*' (Kincaid, 2005, p. 87, italics in the original).

In the second section of this chapter, therefore, we intend to investigate the two pillars of the current social restructuring and technological development of capitalism, namely the role of finance and expansive protection of IPRs, while in the third section we will more specifically consider the efficiency implications of technological innovations and developments taking place within the contemporary capitalist context. The final section will draw the conclusions of this chapter.

7.2 The current restructuring of capitalism, and the role of finance and intellectual property for innovation

7.2.1 Capitalist restructuring and the role of financial capital

In understanding the current process of capitalist restructuring and technological development, Marx's theory of money and credit, scattered in different parts of his work, and the relevant recent developments are perhaps indispensable (see De Brunhoff, 1973; Harvey, 1982, Chs 9–10; Itoh and Lapavitsas, 1999; Lapavitsas, 2003). Contrary to the mainstream theory of money emphasizing its role as a means of exchange, and a Marxist literature stressing the value-form aspect of monetary exchange, a proper value-theoretical Marxist approach to money and credit, as well as the historical evolution of relevant institutions, focuses on the contradiction between money as a means of circulation and money as a measure of value (see Harvey, 1982, p. 249; Kincaid, 2005). Harvey shows, more specifically, that the logic of competition and profitability in both the industrial and financial sectors tends towards a gradual decommodification and undervaluation of money, which threaten to undermine its role as a measure of value. Furthermore, at an international level, he points out that, within the hierarchy of monies, 'the notion of money as a measure of value refuses to die' (Harvey, 1982, p. 295).

Apart from the monetary system and the historical drive towards a decommodification of money, culminating in fiat money, the credit system has also developed historically as an extension of the former, which in various ways facilitates the reproduction and development of capital. There is a complex and contradictory dialectic between real production (extraction of surplus value) and the expansion of monetary and credit means (by the state, the banking or the private sector), determined specifically not only by conjunctural factors associated with the conditions of capitalist accumulation, but also mediated by the definite role of class struggle, ultimately seeking to resist or disrupt the facilitative role of this alien relation (money) which obscures the exploitation of labour. Today, the credit system has expanded beyond the traditional bank credit and the commercial credit between capitalists notably to include the savings and borrowing of working people as well as the organization and promotion of retail sales. Despite this greater complexity, as is aptly pointed out, 'capitalist credit, even in the most developed social forms, is ultimately founded on the ability of industrial capitalists to generate profits through the exploitation of workers. The credit system lends a social and objective aspect to trust and power, but only in order to place both at the service of capitalist exploitation' (Lapavitsas, 2003, p. 69; see also Coates, 2000; Kincaid, 2005).

The close and intimate relation between the real economy and the financial sector particularly needs to be stressed because the rapid expansion and globalization of financial (including speculative) activities have recently led several researchers to erroneously consider it as a largely autonomous and independent development. Pointing to the less rapid and uneven globalization of productive capital, some researchers also erroneously tend to negate or understate the real process of globalization. However, financial capital is merely a particular form of capital (besides productive and commercial capital), closely interconnected with its other forms, and the generalized (and monetized) credit constitutes the most general form of capital. Apart from remaining national disparities or deviations, the modern-day rapid globalization of financial circuits leads to an increasing integration and unification of financial and capital markets. In this case, what is more important is that the terms of financing are unified even for those operations taking place within any domestic context (see Grahl, 2001).

Moving from the monetary and credit relations on a national level to the international level, and the formation of global monetary standards (GMSs), Hampton (2006) suggests that money is constitutive of, and presupposed in, the exploitative and antagonistic relation between labour and its alienated self. The mode of existence of working-class struggle against this exploitative content is a dematerializing money form – a process that weakens the foundations of money's social power. Preserving this power is fundamental to the day-to-day reproduction of capitalist class relations, and capital has sought strategies with which to assert and reinforce this social power by organizing and integrating money at ever higher levels of generalization, culminating finally in GMSs. GMSs are thus institutionalized strategies designed to confront, neutralize or subsume working-class struggle against this alien power.

Contrary to mainstream theory and to a radicalized international political economy (IPE), which simply adds an explanatory discussion of hegemony and international rivalry to the former to end up close to a traditional realist approach, it is further suggested that the rise and fall of GMSs is a tale of the constitution and struggle against money's social power to command living labour (Hampton, 2006).

After the collapse of the Bretton Woods system in the 1970s, chaotic fluctuations of exchange rates and financial deregulation led to a drastic transformation of the sector concerning international finance. Within this context, it becomes clearly apparent that central banks are no longer able to regulate the creation of money, while the principle guiding the emerging global organization and integration of money is the realization of credit as serviced debt: that is, the global intensification of exploitation (Hampton, 2006). Within this still forming and nameless system, while most mainstream or radical analysts are obsessed with the new architecture of international finance and the relevant global governance (see Malhotra, 2002; Hira and Cohn, 2003–4), considerable research focuses on the relation between value theory and international finance. As 'global integration both transcends and at the same time reproduces (national) difference' (Bryan, 2003, p. 60), it is pointed out that increasing international transactions in a variety of financial derivatives, by securing 'commensuration of value across time and space' (ibid. p. 67), bridge this discontinuity and facilitate the transnational extraction and appropriation of surplus value. Within a yet unresolved debate, it is further argued that financial derivatives, having acquired a commodity character, 'perform the role in international finance that gold played in the nineteenth century: they anchor the global financial system' (Bryan and Rafferty, 2006, p. 89). Constituting a representation of abstract labour, they 'thereby create an empirical notion of capital-in-general ... bind the global financial system into an integrated unity in a way that states cannot ... [and can be posited] as an expression of a distinctly global money, operating at a higher level of generality than national currencies' (ibid., pp. 92–3). An objection to this is that '[d]erivatives have no obvious hoarding and paying functions in the world market, and they are certainly not "the anchor of the global financial system". In so far as such an anchor exists today, that is the US dollar' (Lapavitsas, 2006, p. 151).

Under these new conditions, there is ample evidence that industrial investment in developed countries makes little use of funds from banks or other financial institutions such as the stock market, while corporate engagement in the stock market is certainly growing, but mainly in order to finance mergers and acquisitions (M&A) rather than industrial investment. At the same time, corporations are themselves increasingly involved in financial activities, including the provision of credit and trading of financial derivatives (see Itoh and Lapavitsas, 1999; Lapavitsas, 2006).

Finally, it should be noted that, as is commonly realized, the overexpansion and contradictory development of transnational finance, including speculative activity, has greatly increased instability and uncertainty in the world markets. Such instability, however, and the serious financial crises around the end of the last century reflect the deeper crisis and the largely intractable problems of capitalist accumulation (see Kettell, 2005; Nesvetailova, 2005; Lapavitsas, 2006). At the same time, it must be pointed out that the transnational expansion and globalization of finance has a significant impact on the ongoing restructuring of transnational capitalism. It serves in the centralization and reallocation of financial and productive resources at a global level, although it is rather dubious whether this reallocation could ensure a general increase in productive efficiency and, hence, a sustainability of capitalism. It also has definite implications for the rate, content and direction of technological development, but again there is considerable doubt whether these developments constitute a rational change and an improvement for social and human development in general.

7.2.2 The role of finance in technological innovation and development

Contrary to a commonly propagated belief, it has not been the private business sector, but rather public funding (primarily for military programmes) and the role of the state which, historically, have more significantly contributed by far in the development of science and technology (see Noble, 1986; 1995, pp. 96–8; Castells, 2000, p. 69).

Concerning the activities of private firms in R&D and technological innovation, it should be noted that the sources of financing these activities have been drastically differentiated historically. Bank financing and other external sources played a greater role in the past, while recently the internal corporate financing of such activities by retained profits acquires a greater importance. At the same time, there is an apparent shift from public expenditures to private corporate expenditures for research and innovation activities (see Scotchmer, 2004, p. 262). Undoubtedly, private financing and various forms of venture capital operations have contributed towards the advancement of technological innovation, while it is also clear that large firms have greater possibilities, compared to small firms, to finance such technological activities. Of course it is not size alone or market power that matter (see Chapter 12 written by F. Munier on this subject).

The predominant profit imperative of the corporate sector is increasingly leading to the expansion of only those branches of production which are subject to commercial exploitation (commodification), at the expense of the production of public goods, at the same time promoting mainly applied research and related R&D activities, at the expense of basic scientific research (see Nelson, 2004). As pointed out, however, '[l]arge firms may be more willing to undertake basic research when they have a diverse range of products and strong marketing and distribution networks that increase their confidence [so] that they will eventually be able to put the findings of basic research to some good commercial use' (Rosenberg, 1990, p. 168). It is important to note though that, even under such conditions, the deepening capitalist crisis and the intensification of competition tend to reduce private basic research.

The previous point is closely related to the increasing concern of corporate capital for the appropriability of innovations stemming from R&D activities. Great attempts are made to protect, in various ways, the innovations and knowledge deriving from such R&D activities, and hence financial investment in those activities is restricted by those conditions of appropriability. The expanding protection of intellectual property during recent decades (mainly through patents), to be considered in more detail in the next section, is a particular case of increasing importance. According to available evidence, however, patent protection seems to be effective against imitation only in a restricted number of industries (mainly those involved with chemicals and pharmaceuticals), while other forms of protection, such as secrecy, lead time, and sales and service efforts, are more effective for the protection of process and product innovations in most industries (see Harabi, 1995; Scotchmer, 2004, pp. 259, 261).

Under these conditions, the increasing significance of science and the rapid commercialization of knowledge have increasingly made research and technological knowledge into a partly autonomous (with a large number of offspring and related venture capital firms), very profitable, and rapidly expanding enterprise. Aside from other mechanisms for the transnational development and transfer of technology, there is evidence that, although most corporations seem to retain most of their technological activities in their home base, a rising part of such activities is transnationalized. The greatest part of transnational relocation of such activities concerns the core areas where corporations are strong at home, and the motives for this transfer are mainly related to the need to adapt products, processes and materials to suit foreign markets, as well as to provide relevant technical support to offshore manufacturing plants (see Patel and Vega, 1999). The overall development of transnational finance undoubtedly raises the potential of R&D financing and facilitates a similarly transnational development and application of new technologies. At the same time, it unifies the pattern of technological development and determines more tightly its class character.

The rising significance of science and the increasing commodification of technical knowledge, partly a consequence of capitalist financing, have developed in parallel with the entrepreneurialization of universities and their increasing interconnectedness with the corporate business sector (one aspect of innovation networks, studied by D. Uzunidis in Chapter 2). The process of entrepreneurialization of universities and other research centres, which is both an exogenous and endogenous process (see Etzkowitz, 2003; Scotchmer, 2004, pp. 235–40; cf. Johnson et al.,

2003), does not simply imply a private financing of university research, as it gives big corporations the potential to extract huge social resources by freely appropriating scientific knowledge and technological innovations developed in a university sector largely financed by public funds. What is more, these new evolutions are combined with a predominance of corporate profitability criteria, hence implying an enormous misdirection of scientific and social research. Under these conditions, the older dilemma, between a publicly financed research sector under the guidance of the state so as to meet research priorities and social needs, and a similarly self-governed publicly funded and autonomous 'Republic of Science', is largely outmoded. This process of privatization of research and scientific knowledge by leading to the shrinkage of scientific commons implies further 'bad news for both the future progress of science, and for technological progress' (Nelson, 2004, p. 455).

Apart from serving the profitability of capital and its tighter control, the specifically capitalist pattern of financing research and technological innovation has serious negative implications for the rate, content and direction of technological development (see also Scotchmer, 2004, p. 27). The short-sighted prioritization of applied research and the monopolistic protection of intellectual property imply a decline in the rate of innovations, thus retarding scientific and technical progress and the overall development of the social forces of production. The capitalist criteria of financing and technological development imply a misdirection of research and innovations away from real social needs, while the specific technology developed tends to be alienating and exploitative for labour, and often destructive for the ecological environment (see Noble, 1995; McMurtry, 2002, pp. 115–17, 137; Liodakis, 2003; UNESCO, 2005, pp. 103, 117, 169). Moreover, while new technologies (especially ICTs) contribute to an increasing socialization of production as well as the technological development enterprise, the expanding privatization of research and the commodification of knowledge tend to undervalue or marginalize local and indigenous knowledge, eliminate the priceless heritage of humanity, and thus reduce cognitive diversity as a source of social sustainability. At the same time, they tend to seriously restrict the potential for technology and knowledge sharing, thus restricting the potential for productive efficiency and social sustainability (see Nelson, 2004; UNESCO, 2005, pp. 22–3, 48, 58). Finally, it should be noted that a sufficient and successful financing of (product and process) innovations may increase productivity and create the potential for surplus profits in a particular industry or region, which in turn can finance further R&D, leading to a virtuous circle of productivity increase and development. In this way, the rapidly expanding financial capital tends to reinforce the inherent tendency of industrial capitalism towards uneven development. Today, there is ample evidence of an extremely uneven technological development across both countries and industries (see OECD, 2005). Taking the other side of the dialectic, we should also point out that the development of technology (and new technologies in particular), not only enables the development and globalization of finance, but also contributes to a potential improvement in the valorization conditions of capital, and hence to the specific terms of credit monetization and the development of financial capital on a global scale.

7.2.3 The impact of an expansive protection of intellectual property on technological innovation and development

Although the attempt to monopolize knowledge and intellectual property is as old as capitalism itself, the huge surge towards an expansive protection of IPRs during recent decades reached its culmination with the TRIPs Agreement of the WTO (1994) and the associated global regulation and institutionalization of the protection of IPRs (see Correa, 2000; Richards, 2004; Drachos and Braithwaite, 2004; May, 2006). The reasons for this profound institutional restructuring of capitalism should be sought in two directions. On the one hand, it is the growing significance of scientific knowledge, associated in particular with the recent revolutionary developments in technology (new technologies), which creates the potential as well as the need for capital to monopolistically protect relevant IPRs. On the other hand, the deep accumulation crisis of capitalism creates the pressing need as a move forward for an expansion and deepening of the capitalist relations of production, which can be partly realized through such a monopolistic protection of IPRs. This interpretation is at variance with conventional institutional functionalism, which considers this institutional and legal restructuring, in a naturalistic and reified fashion, as a necessity serving the further development of technology and knowledge.

According to the capitalist logic underpinning this process of IPRs protection, it purports to make investment in research and technological innovation profitable, and thus give further incentives for undertaking R&D expenditures and for a technological development which fuels further economic growth. Undoubtedly, the capitalist criteria of financing such activities have significantly induced a further commercialization of knowledge and monopolistic protection of IPRs. The experience during recent decades, however, indicates that this ever-expanding protection of IPRs has not brought about the purported results for technological innovation and growth (see Jaffe, 2000; Kumar, 2003). Even within conventional innovation economics, considerable doubt is expressed about the particular forms of protection, as well as increasing concerns about the duration and scope of patent protection, the entrepreneurialization of universities and their exclusive licensing of technological innovations, the stifling of technological and knowledge spillovers, the shrinkage of scientific commons, and the overall decline in the efficiency of the technological innovation effort (see Harabi, 1995; Mazzoleni and Nelson, 1998; Jaffe, 2000; Nelson, 2004). As stated, despite the overwhelming support of this reform, '[t]he world economy will not benefit from a general broadening and strengthening of patent rights' (Mazzoleni and Nelson, 1998, p. 281).

From a political economy standpoint, a critique of the reification of IPRs, by uncovering the naturalization, depoliticization and technocratic rationalization of these reforms, tends to historicize IPRs and explicitly recognize that it is the interests of capital and its historically specific needs, under current conditions, that determine this move towards an expansive protection. As pointed out, at the same time, 'one of the key functions of IPRs is to construct a scarcity when none necessarily (nor naturally) exists, in order to support price-making and the subjection of knowledge to capitalist economic relations' (May, 2006, p. 48). As already noted, more fundamentally, the expansion of IPRs serves as a new type of 'intellectual enclosures' in promoting inequalities and the expansion of the market relations of production.

Apart from the creation of scarcity, the reinforcement of economic power concentration and the expansion of the markets, there is an extensive literature concerning the negative implications of IPRs, such as the retardation of technological progress and the development of productive forces, the misdirection of research, failing to serve real social needs, its exploitative connotations against labour and the ecologically destructive character of the technology developed, and the 'monstrous effects' on higher education (see Perelman, 1998, pp. 52–3; 2003; May, 2002; McMurtry, 2002, Ch. 4; Liodakis, 2003, 2006; Drachos and Braithwaite, 2004; UNESCO, 2005, pp. 116–17, 142–46). There is also a growing concern about the negative impact of IPRs and the TRIPs Agreement on developing countries, and more specifically about 'choking an important contributor of growth that has been variously described as imitative duplication, reverse engineering or knowledge spillovers from abroad' (Kumar, 2003, p. 222; see also Correa, 2000).

A re-evaluation of the experience of recent decades concerning IPRs increasingly leads even mainstream economists to underline the urgent need for a drastic reform of the IPRs regime (see Mazzoleni and Nelson, 1998; Jaffe, 2000; Gallini, 2002; Nelson, 2004; Baker, 2005). What is more at stake, however, is the specific mode of utilization of the available means of production (including scientific and technical knowledge) and the specific pattern of financing and technological development. In this sense, the growing unevenness and class-determined character of technology, and the misguided, ecologically degenerating and dehumanizing character of the dominant pattern of economic and technological development raise serious questions and open new discussions concerning both alternative property regimes (see Runge, 1986; Ostrom, 1990; Agrawal, 2001), and alternative models of production and utilization of technology. This latter orientation of research points to the need for more cooperative production and sharing of technology, such as the Open Source software, and includes both a mainstream perspective (see Lerner and Tirole, 2005; UNESCO, 2005, pp. 159, 169–78), and a more radical perspective (see Barbrook, 2000; Bauwens, 2006) which may extend to more fundamental questioning and call for fundamental institutional change. These changes are linked to the increasingly social character of production (technology and knowledge) and its currently expanded private appropriation (see Liodakis, 2006).

As the apparently flawed character and the growing inefficiency of the technological enterprise make the sustainability of capitalism itself problematic, there is an urgent need to more specifically question the concept itself of productiveness and efficiency as it is commonly used and meant in mainstream parlance.

7.3 Technological innovation and the myth of capitalist efficiency

The dominant capitalist ideology includes a myth, which stresses a putative allocative efficiency of the market mechanism, and relates it to competition and an overstressed technological dynamism of capitalist production, leading to continuous cost-reducing technological innovation. However, this dogmatic mythology is based on a clearly inadequate conception of both *productivity* and *economic efficiency*. Productivity is usually conceived in physical terms as a technical relation between inputs and outputs of production (input minimization/output maximization), and this conception is associated with market efficiency (at a micro level), coupled at a more general (macro) level with the so-called Pareto optimality and the concept of *Pareto efficiency*. But this latter concept, apart from its metaphysical abstractness, is again theoretically problematic insofar as it is irrelevant in the most general case, while in the particular case (when no one can become better off without someone else becoming worse off), it is theoretically inadequate, socially biased and narrow-minded. This is because it tacitly assumes an unchangeable structure of property rights and obscures the initial distribution of property (status quo). This theoretical inadequacy leads to the paradox that it is perfectly possible for a particular allocation of resources, or a specific socio-technical arrangement, to be 'efficient' and at the same time not capable of adequately (or not at all) meeting even the basic needs of many citizens.

A critical interrogation of this mythology about capitalist efficiency should start by questioning the productivity concept in physical terms (see Harvey, 1982, pp. 102–6). As Marx stressed, the productivity of labour, and we can meaningfully speak only of the productivity of labour (and *not* of capital), is in these terms and in its apparent generality seriously misleading. As the social relations of production involved and the overriding ends (the purpose) of the prevailing mode of production are crucial, we can come up with a meaningful measure of productivity only by taking these conditions into account. In this sense, the essence of productivity associated with the satisfaction of human needs is very different from productivity in relation to the creation of surplus value. Under capitalist conditions of production, Marx argued that productive labour is only that labour which produces surplus value for capital (1967, I, p. 509). Productivity, therefore, should be specifically understood in *value* rather than *physical* terms. Moreover, what is important is not so much individual labour productivity, but rather the collective productivity of labour. In this regard, it should also be understood that economic (or social) efficiency is not simply a technical issue insofar as social relations and the purposes of production can condition and decisively determine its content.

It should further be noted that, within a historically specific context, market failures (serious external effects) may imply a misallocation of resources, a suboptimal structure of production, and a serious impact on (the rate, content and direction of) technology, which may result in a serious decline in overall social efficiency (see Klibanoff and Morduch, 1995; Schmitz, 1999; May, 2002; McMurtry, 2002, pp. 139, 151, 153). This is apparently the case when negative externalities are not taken into account (cost shifting), and positive externalities are insufficiently utilized due to the fragmented and (individually) antagonistic character of production. In this case the efficiency of a single firm, or a number of firms, does not imply the general efficiency of the system. On the contrary, the contemporary conditions of a protracted crisis and intensified competition imply a tendency for increasing externalities (market failures), and hence for a declining efficiency and stifling of the buoyancy of capitalist production. Under these conditions, the content itself and the direction of modern technologies tend also to undermine the fundamental resources of production (labour and nature) and further reduce the efficiency of capitalist production (see McMurtry, 2002, pp. 101, 107–9; Liodakis, 2006).

It is also remarkable that, although some mainstream economists recognize that collective efficiency may derive from (positive) external economies and conscious cooperation (joint action), trapped within the existing conditions and the prevailing methodological individualism, they fail to see the extension of conscious collective action in a social planning which might raise efficiency and sufficiently meet social needs (see Schmitz, 1999).

Undoubtedly, technology continues, in the current stage of capitalism, to constitute a basic means for advancing capital reproduction and accumulation. In some cases it presents a particular dynamism. Biotechnology contributes, among other things, to a drastic reduction in the cost of labour reproduction (see Liodakis, 2003), while ICTs, by speeding up circulation, tend to improve the valorization conditions of capital. However, the inefficiency of capitalist production in general, and of the process of technological development in particular, seems to be growing, as analysed above, and technological innovation appears insufficient or incapable of busting capitalist accumulation and ensuring sustainability of the CMP. The continuing and exacerbating crisis, apparently, calls for a drastic reorientation of the process of technological development, along with a perhaps more fundamental social restructuring.

7.4 Social restructuring and the need for a reorientation of technological innovation

In our investigation of the dialectic between current socio-economic restructuring and technological development, we have focused particularly on two aspects (the main pillars), namely the transnational development of financial capital and the expansive protection of IPRs, which seem to be of specific significance for the determination of technological innovation and development, within the fundamental restructuring associated with the currently constituted transnational (totalitarian) capitalism. The process itself of technological innovation and development has been considered as a set of techniques, organizational forms, and scientific/technical knowledge which, due to their specific class content and orientation, determined (among others) by the above-mentioned factors (capitalist financing and IPRs), are put to the service not of satisfying human needs, but rather of a further command and exploitation of human and other resources, purporting a maximal accumulation of profits for capital.

In this new stage of capitalism, private property, not only in physical assets and means of production but also in scientific and technical knowledge, despite its inherent undermining and transformation, maintains its crucial importance. Although a considerable development of new access relations over a series of means of production is currently taking place, these developments do not really question private property in its most general sense. Although the evolution of private property institutions, under the new social and technological conditions, raises significant issues, concerning alternative property regimes and the sustainability itself of capitalism, it is not property relations themselves but rather the capitalist relations of production which are most significant, namely the specific mode of production, extraction and appropriation of surplus value, and the historically specific terms under which this takes place. The capitalist relations of production could be reproduced and developed even on the basis of a system of extensive collective (or public) property and access relations.

As stressed moreover, the flawed character and the misguided orientation of the process of technological development exacerbate the problem of a growing inefficiency in capitalist production; they also pose the need for a reorientation of technological innovation. This reorientation is obviously meant in the sense of a greater democratic control of the process of technological innovation and development itself, which places this development in the service of satisfying human needs rather than the profit imperative and accumulation of capital.

This glaring need for a reorientation of technological development reflects an intensifying contradiction between the rapidly developing social forces of production and the particular structure of the social relations of production characteristic of modern capitalism. As the current trends in finance, property relations (including IPRs) and technological development tend to undermine and seriously endanger scientific commons, there is obviously an urgent need for further research and the development of a radically different management of the process of technological innovation, which would both resolve the above-mentioned contradiction and open new perspectives for social restructuring and human development.

References

Agrawal, A., 'Common Property Institutions and Sustainable Governance of Resources', *World Development*, 29 (10) (2001), 1649–72.

Baker, D., 'The Reform of Intellectual Property', *Post-Autistic Economic Review*, 32 (2005).

Barbrook, R., 'Cyber-Communism: How the Americans are Superseding Capitalism in Cyberspace', *Science as Culture*, 9 (1) (2000), 5–40.

Bauwens, M., 'The Political Economy of Peer Production', *Post-Autistic Economics Review*, 37 (2006), 33–44.

Braverman, H., *Labor and Monopoly Capital: the Degradation of Work in the Twentieth Century* (New York: Monthly Review Press, 1974).

Bryan, D., 'Bridging Differences: Value Theory, International Finance and the Construction of Global Capital', in R. Westra and A. Zuege (eds), *Value and the World Economy Today: Production, Finance and Globalization* (London: Palgrave Macmillan, 2003).

Bryan, D. and M. Rafferty (2006) 'Money in Capitalism or Capitalist Money?', *Historical Materialism*, 14 (1) (2006), 75–95.

Castells, M., *The Rise of the Network Society* (London: Blackwell, 2000).

Carchedi, G., 'On the Production of Knowledge', in P. Zarembka (ed.), *The Capitalist State and Its Economy; Democracy in Socialism*, Series *Research in Political Economy*, 22 (2005), 261–98.

Coates, D., *Capitalist Models and Their Discontents* (Basingstoke: Palgrave Macmillan, 2000).

Correa, C., *Intellectual Property Rights, the WTO and Developing Countries: the TRIPS Agreement and Policy Options* (London: Zed Books, 2000).

De Brunhoff, S., *Marx on Money* (New York: Urizen, 1973).

Drachos, P. and J. Braithwaite, 'Who Owns the Knowledge Economy? Political Organising behind TRIPS', The Corner House, Briefing 32 (2004).

Etzkowitz, H., 'Research Groups as "Quasi-firms": the Invention of the Entrepreneurial University', *Research Policy*, 32 (1) (2003), 109–21.

Gallini, N., 'The Economics of Patents: Lessons from Recent U.S. Patent Reform', *Journal of Economic Perspectives*, 16 (2) (2002), 131–54.

Grahl, J., 'Globalized Finance', *New Left Review*, 8 (2001).

Hampton, M., 'Hegemony, Class Struggle and the Radical Historiography of Global Monetary Standards', *Capital & Class*, 89 (2006).

Harabi, N., 'Appropriability of Technological Innovations', *Research Policy*, 24 (1995), 981–92.

Harvey, D., *The Limits to Capital* (Oxford: Basil Blackwell, 1982).

Hira, A. and T. Cohn, 'Toward a Theory of Global Regime Governance', *International Journal of Political Economy*, 33 (4) (2003–4), 4–27.

Itoh, M. and C. Lapavitsas, *Political Economy of Money and Finance* (London: Macmillan, 1999).

Jaffe, A. B., 'The U.S. Patent System in Transition: Policy Innovation and the Innovation Process', *Research Policy*, 29 (4–5) (2000), 531–57.

Johnson, B., P. Kavanagh and K. Mattson (eds), *Steal this University: the Rise of Corporate University and the Academic Movement* (New York and London: Routledge, 2003).

Kettell, S., 'Circuits of Capital and Overproduction: a Marxist Analysis of the Present World Economic Crisis', *Review of Radical Political Economics*, 38 (1) (2005), 24–44.

Kincaid, J., 'A Critique of Value-Form Marxism', *Historical Materialism*, 13 (2) (2005), 85–119.

Klibanoff, P. and J. Morduch, 'Decentralization, Externalities, and Efficiency', *Review of Economic Studies*, 62 (2) (1995), 223–47.

Kumar, N., 'Intellectual Property Rights, Technology and Economic Development: Experiences of Asian Countries', *Economic and Political Weekly*, 38 (3) (2003), 209–25.

Lapavitsas, C., *Social Foundations of Markets, Money and Credit* (London: Routledge, 2003).

Lapavitsas, C., 'Relations of Power and Trust in Contemporary Finance', *Historical Materialism*, 14 (1) (2006), 129–54.

Lerner, J. and J. Tirole, 'The Economics of Technology Sharing: Open Source and Beyond', *Journal of Economic Perspectives*, 19 (2) (2005), 99–120.

Liodakis, G., 'The Role of Biotechnology in the Agro-food System and the Socialist Horizon', *Historical Materialism*, 11 (1) (2003), 37–74.

Liodakis, G., 'The New Stage of Capitalist Development and the Prospects of Globalization', *Science & Society*, 69 (3) (2005), 341–66.

Liodakis, G., 'The Global Restructuring of Capitalism, New Technologies and Intellectual Property', in B. Laperche, J. K. Galbraith and D. Uzunidis (eds), *Innovation, Evolution and Economic Change: New Ideas in the Tradition of Galbraith* (Cheltenham, UK: Edward Elgar, 2006).

Malhotra, K., 'Renewing the Governance of the Global Economy', in W. Bello, N. Bullard and K. Malhotra (eds), *Global Finance: New Thinking on Regulating Speculative Capital Markets* (London and New York: Zed Books, 2000).

Marx, K., *Capital* (New York: International Publishers, 1967).

Marx, K., *Grundrisse* (London: Penguin Books, 1973).

May, C., 'Unacceptable Costs: the Consequences of Making Knowledge Property in a Global Society', *Global Society*, 16 (2) (2002), 123–44.

May, C., 'The Denial of History: Reification, Intellectual Property Rights and the Lessons of the Past', *Capital & Class*, 88 (2006).

McMurtry, J., *Value Wars: the Global Market versus the Life Economy* (London: Pluto Press, 2002).

Mazzoleni, R. and R. R. Nelson, 'The Benefits and Costs of Strong Patent Protection: a Contribution to the Current Debate', *Research Policy*, 27 (1998), 273–84.

Nelson, R., 'The Market Economy, and the Scientific Commons', *Research Policy*, 33 (3) (2004), 455–71.

Nesvetailova, A., 'Fictitious Capital, Real Debts: Systemic Illiquidity in the Financial Crises of the Late 1990s', *Review of Radical Political Economics*, 38 (1) (2005), 45–70.

Noble, D., *America by Design: Science, Technology, and the Rise of Corporate Capitalism* (Oxford and New York: Oxford University Press, 1977).

Noble, D., *Forces of Production: a Social History of Industrial Automation* (New York: Oxford University Press, 1986).

Noble, D., *Progress without People: New Technology, Unemployment, and the Message of Resistance* (Toronto: Between the Lines, 1995).

Noble, D., 'Digital Diploma Mills: the Automation of Higher Education', *Science as Culture*, 7 (3) (1998), 355–68.
OECD, *Science, Technology and Industry Scoreboard* (Paris: OECD, 2005).
Ostrom, E., *Governing the Commons: the Evolution of Institutions for Collective Action* (Cambridge: Cambridge University Press, 1990).
Patel, P. and M. Vega, 'Patterns of Internationalization of Corporate Technology: Location vs. Home Country Advantages', *Research Policy*, 28 (2) (1999), 145–55.
Perelman, M., *Class Warfare in the Information Age* (London: Macmillan, 1998).
Perelman, M., 'Intellectual Property Rights and the Commodity Form: New Dimensions in the Legislated Transfer of Surplus Value', *Review of Radical Political Economics*, 35 (3) (2003), 304–11.
Richards, D., *Intellectual Property Rights and Global Capitalism: the Political Economy of the TRIPs Agreement* (Armonk, NY: M.E. Sharpe, 2004).
Rifkin, J., *The Age of Access* (New York: J. Tarcher/Putnam, 2000).
Rosenberg, N., 'Why do Firms do Basic Research (with Their Own Money)?', *Research Policy*, 19 (2) (1990), 165–74.
Runge, C. F., 'Common Property and Collective Action in Economic Development', *World Development*, 14 (5) (1986), 623–35.
Schmitz, H., 'Collective Efficiency and Increasing Returns', *Cambridge Journal of Economics*, 23 (1999), 465–83.
Scotchmer, S., *Innovation and Incentives* (Cambridge, Mass.: MIT Press, 2004).
UNESCO, *UNESCO World Report: Towards Knowledge Societies* (Paris: UNESCO Publishing, 2005).

Part II
Innovation Trajectories and Profitability in Firms' Strategies

8
The Firm and its Governance over the Industry Life Cycle

Jackie Krafft and Jacques-Laurent Ravix

Uniformity in modes of governance of the firm is now widely debated. So far, the predominant thesis was that there should be a superior model promoting optimality by disclosure of information and transparency. But today this thesis is greatly contested, since the adoption of a unique and universal set of rules and arrangements neglects the diversity and heterogeneity of firms, industries, as well as institutional contexts (Becht et al., 2005). Moreover, evidence shows that this unique model of governance tends to generate major failures and turbulence, especially in innovative industries (Fransman, 2002; Lazonick and O'Sullivan, 2002; Krafft and Ravix, 2005). What emerges as a result is that different types of rules and norms should govern differently entrepreneurial as well as public firms, depending on the industry in which they operate and the stage of development of this industry.

This chapter explores this issue by reconciling two trends of literature that are generally disconnected – the industry life cycle (ILC) on the one hand and the governance of large and small firms on the other – to generate results on how the governance of the firm may look like over the industry life cycle. When the two bodies of literature are connected, the immediate result is that the governance of small, young and innovative firms in the early stages of the life cycle should be different from the governance of large, mature and routinized firms. Small young and innovative firms should benefit from a mode of governance based on *cooperation and assistance* to stimulate innovation, while large mature and routinized firms should be imposed a mode of governance based on *control* of the manager's action in the interests of shareholders. We argue that this immediate result can only but be preliminary, since age and size are not necessarily the key determinants of innovative behaviours of firms. In the ILC, small new firms engage in product innovations, while large mature firms continue the process of innovation by investing in process capacities.[1] In that perspective, imposing on these firms a governance based on control may not be the optimal solution, since we know that this mode of governance favours short-term choices that may be detrimental to the development of innovation. What is more important is thus to consider how the innovative behaviour of firms can be maintained in phases of growth and decline of the industry. In this chapter, we advance the idea that new principles of governance should be proposed for innovative corporations (large or small) as a distinctive category.

As a first step, we review the literature on the ILC and on the governance of small and larger firms. We discuss the implications in terms of governance in the early stages of the life cycle and in the later stages. We argue that the vision of governance that results is too strongly based on the assumption that firms are highly innovative at the beginning of their life and much less as they age. This is not necessarily coherent with the ILC in which firms are innovative all along their life, and further questions the principle that some of them should be governed through a cooperation and assistance mode, while others should be governed by a control mode. In a second step, we propose that a more appropriate vision is to consider that firms, independently of their age and size, may be involved in radical innovation processes and, all along the development of such processes, have to face the competition of rival firms engaged in predator strategies. When innovation is put at the centre stage of the analysis, it is possible to show on the basis of an evolutionary game that, for a large range of parameters, the innovative strategy tends to be dominated over time by the predator strategy, if no external forces, such as corporate governance preserving long-term innovation projects, emerge. In a final step, we propose thus that the real determinant of assigning different modes of governance should be the presence of innovation, suggesting that cooperation and assistance have to be the key reference in that case, while the absence of innovation could alternatively legitimate control based on shareholder value maximization as the leading principle. We derive new perspectives on the governance of innovative corporations, by defining the notion of 'corporate entrepreneurship' within which managers and investors are collectively involved in the coherence and development of small, but also large innovative firms.

8.1 Industry life cycle and the governance of firms

Since the original paper by Gort and Klepper (1982), it is now common knowledge among economists that key features of firms change as they age and progress over the life cycle. More recent contributions on the theme (Klepper, 1997) show that, in the early stages of the ILC, many firms are product innovators, most of them are profitable and very few exit the industry. They operate on a small-scale basis, each firm representing very small market shares. In the final stages of the cycle, in contrast, firms tend to be big process innovators. They are very few in number and have large market shares. What is much less debated in the ILC literature is whether the governance of these firms changes over the ILC and how. Should firms be owned and managed the same way at the time they emerge, grow, age and decline? Or should there be distinct types of corporate governance along the phases of the ILC? On the one hand, the literature on start-ups and venture capital suggests that firms should be governed on the basis of a close cooperation between the founder entrepreneur (or professional manager) and the investor (business angel, venture capitalist). On the other hand, the literature on the governance of corporate firms generally supports the shareholder value vision in which the relationship between the manager and the investor is in terms of conflicting objectives, leading to a realignment of the manager's incentives in the investor's interests. The conclusion of these two trends of literature is thus that there should be distinct modes of governance over the ILC, one dedicated to small, young and risky firms, and based on cooperation; the other dedicated to older and mature firms, and based on control.

8.1.1 Key stylized facts of an ILC

The ILC literature proceeds from a basic biological analogy, positing that industries, like biological organisms, have different periods in their life (birth, growth, maturity, decline and death) and that their key characteristics change over time. In the following, we present major features of this body of literature, distinguishing what occurs in the early and late phases of the ILC.[2]

8.1.1.1 The early stages of an ILC

The early stages of an ILC are composed of phases 1 and 2, namely emergence and growth phases. Key stylized facts for innovation and organization of the industry are the following:

- *Innovation*: innovation is imported from technologically related industries, and diffused in the emerging industry by exogenous information. Innovation generally concerns product definition and improvement, and generates an increase in product variety.

- *Organization of industry*: production increases, and since opportunities for profit are important, entry increases also. The volatility of market shares is very high.

8.1.1.2 The late stages of an ILC

The late stages of an ILC are composed of phases 3 to 5, namely maturity, decline and death phases. Key stylized facts for innovation and organization of the industry are the following:

- *Innovation*: innovation is endogenously created by the experience of the incumbents, which may create barriers to entry for newcomers. Innovation concerns the process of production. Dominant designs tend to be adopted by end-users and standardization phenomena occur at the level of producers.
- *Organization of industry*: production falls, and a massive exit of firms occurs (shake-out). Market shares tend to stabilize and first movers benefit from a competitive advantage.

8.1.2 Governance in the early stages of the ILC: cooperative governance of start-ups

The ILC gives the small new firms a key role in the impulsion of innovation. They are at the origins of a life cycle. This vision is corroborated by other approaches in industrial dynamics that focus on the asymmetric size distribution of firms, with a small number of large companies and a large number of small firms. This skewed firm size distribution has a remarkable persistence across industries, countries, and over time (Geroski, 1995). These numerous, small, new firms are moreover seen as crucial to economic development, especially because they are generally at the origins of new technological and market opportunities whereas older incumbent firms are in a phase of decline of the life cycle (Audretsch, 1995).

In that perspective, the question of how these firms are financed during the seed phase is a key issue. Gompers and Lerner (2001) argue that a venture capital revolution has emerged for these firms. They sustain that (ibid., p. 145): 'Venture capital is now an important intermediary in financial markets, by providing capital to firms that might otherwise have difficulty attracting financing. These firms are typically small and young, plagued with high levels of uncertainty and large differences between what entrepreneurs and investors know.'

In the literature, this issue on the respective knowledge of the manager and the investor has been treated for a long time in agency terms, i.e. in a framework based on asymmetric information and complete contracts. The entrepreneur has big incentives to engage in unproductive expenditures, since he does not bear their entire cost; or to develop an insufficient level of effort, since this level is not directly observable by the investor. These important information asymmetries between the entrepreneur and the venture capitalist can be solved on the basis of a complete or quasi-complete contract (Jensen and Meckling, 1976; Grossman and Hart, 1986; Hart and Moore, 1988). The solution broadly lies in the investor's scrutinization of firms before providing capital and monitoring them afterwards. The outcome is, very often, highly complex venture capital contracts (Gompers, 1995, 1996; Kaplan and Stroemberg, 2003, 2004) that limit their applicability in the real world. In addition to the argument on the lack of simplicity, Aghion and Bolton (1992) show that inefficiency not only affects managers, and that unsolvable agency problems (involving the manager and the investor) may arise. Because of uncertainty, situations arise that cannot be foreseen or planned for in an initial contract.

New developments thus tend to recognize that the relation between the investor and the manager is necessarily based on incomplete contracts (Audretsch and Lehman, 2006). In that case, what entrepreneurs and investors know is highly dependent on their specific skills, experiences and practices. Since this knowledge is not easily transferable, the investor and the manager have to develop close connections in order to progressively share their respective knowledge. Close connection is especially necessary, since lenders have to deal with evaluating innovative but less proven business concepts. Small new firms do not generally demonstrate an established history of earning and financial stability. Also, for many start-ups, the primary assets are intangible and difficult to value, thus failing to satisfy requirements for asset-based security. In that case, venture capitalists and business angels finance new and rapidly growing companies, and especially purchase equity securities. But, to do this, they generally assist the development of new products or services, and add value to the company through active participation. They usually take higher risks with the expectation of higher rewards, and have a long-term orientation.

The nature of the relationship between the manager and the investor is thus based on cooperation and assistance: the founder-entrepreneur or the professional manager has to diffuse his own knowledge on the characteristics of his innovation and market potentialities, while the business angel or venture capitalist has to propose different solutions to finance the initial step of elaboration of the innovative project, as well as its development over time.

8.1.3 Governance in the late stages of the ILC: control-based governance of mature firms

The ILC views large firms as key actors in the development of innovation, especially by their greater capacity to invest in process innovation, based on the accumulation of knowledge and competences since their entry at the beginning of the life cycle. However, this phenomenon can also be analysed in reference to the pervasive effects it may generate, such as the erection of barriers that deter innovative entry, the dominance of suboptimal dominant designs and standards, and eventually the engagement in inefficient choices from the manager in a situation where large size increases bureaucracy and decreases the intensity of competition. These pervasive effects, and especially the third one, are at the core of the literature on the governance of large, mature firms.

In the big corporation, the governance problem is essentially to persuade the manager to behave fairly on behalf of the investor, and to avoid any discretionary behaviour. The general solution to this agency problem is to grant managers a highly contingent, long-term incentive contract *ex ante* to align his interests with those of his principals (Schleifer and Vishny, 1997). The formalization, strongly based on a complete contract hypothesis, provides the essential requirements of corporate governance oriented towards shareholder value within a context of transparency of information and generalization of contractual relations in organizations. Managerial corrections may take various forms (board of directors, proxy fights, hostile takeovers, corporate financial structure), and are always oriented towards monitoring and disciplining management in the interest of shareholders and investors.

Complementary approaches are also developed on the basis of transaction costs (Williamson, 1985), and property rights (Hart, 1995a) in order to consider weaker rationality hypotheses, and higher costs of negotiating and writing down contracts. This literature more deeply relies on notions of incomplete contracts and residual rights of control[3] that are absent in agency theory. But, despite these differences, transaction costs and property rights literature generally come to the same conclusions as agency theory concerning the rules of governance of large publicly held companies (Hart, 1995b; Williamson, 1988, 2000).

The nature of the relationship between the manager and the investor is based on control: the investor orientates and monitors the choices of the manager. The investor, from the information of key indicators such as return on investment, or economic value added/market value added, has the capacity to evaluate whether the manager has behaved fairly to shareholders or not. From these indicators, the investor checks whether the manager has transformed his background knowledge into shareholder value-maximizing strategies.

8.1.4 Summing up

When we relate the literature on the ILC with that on the governance of firms, we end up with two sets of results: one related to the governance of firms in the early stages, where cooperation and assistance modes of governance should dominate, and one related to governance in the late stages that should be based on control and realignment of incentives. In small new firms that operate in the early stages of the ILC, the manager is the innovator, the founder entrepreneur or a professional manager, whose role is to discover new technological and market opportunities. The investor is often a business angel or a venture capitalist who assists in the development of new products, adds value to the company, takes higher risks with the expectation of higher rewards, and has a long-term orientation. The governance is thus based on cooperation and assistance, and is supported by different structures, such as the development of scientific and R&D committees, to increase the long-term performance of the firm. Alternatively, large and mature firms that dominate the late stages of development of the industry are governed by a board of directors that run the company in the interests of the shareholders. Very often these shareholders are institutional investors, such as pension funds, that tend to realign the manager's incentives, fight against the manager's discretion, and assess the value of the company on purely financial criteria. They are short-term and risk minimization oriented (Table 8.1).

The issue now is whether we can consider this dichotomy as robust in the main changes that occur all along the ILC, and especially in the development of innovation. Firstly, we can note that this dichotomy is mainly driven by a specific vision of firms, being very innovative at the beginning of their life and much less as they age. This vision can be discussed, especially since the ILC does not necessarily end up with these drastic conclusions. Firms, as they age, tend to reduce their spectrum of product innovation, but are the sole firms to possibly invest in process innovation. Moreover, first movers that become the leaders of the industry build their competitive advantage step by step, since what they do in each period of the ILC has direct implications on subsequent periods. Finally, barriers to entry are related to the exploitation of knowledge

138 Powerful Finance and Innovation Trends

Table 8.1 Governance in the early and late stages of the ILC

	Early stages: emergence and growth of the industry	Late stages: maturity and decline of the industry
Type of firm	Small, new firms	Large, mature firms
Type of manager	Innovator, founder entrepreneur, professional manager	Board of directors
Role of manager	Discover new technological and market opportunities	Run the company in the interest of the shareholders
Type of investor	Business angel, venture capitalist	Institutional investors (pension funds)
Role of investor	Assist the development of new products or services Add value to the company through active participation Take higher risks with the expectation of higher rewards Have a long-term orientation	Realign the manager's incentives Fight against manager's discretion Assess the value of the company on the basis of financial criteria Minimize risks Have a short-term orientation
Nature of governance	Cooperation and assistance	Control
Structure of governance	Scientific committees Research and development committees	Audit committees, compensation systems, proxy fights, hostile takeovers

and experience accumulated over time by firms, and not necessarily to the willingness of incumbents to deter entry. Secondly, we can also note that ideas on the governance of small firms have greatly changed over time, starting from basic control modes of governance inspired by agency theory, and ending up with more operational and pragmatic modes of governance based on cooperation and assistance. On this point, the recognition that the manager's knowledge was necessarily different from – yet highly complementary to – that of the investor, has been determinant in the change of vision. We think that a similar argument should also be investigated at the level of the large innovative firm.

8.2 Innovative behaviours and predator constraints

In what follows, we consider that innovation, which is at the centre stage of the ILC literature, should be a major criterion for defining the appropriate modes of governance over time. In fact, because innovation can either be long or short term oriented, different modes of governance might be required for firms engaging in these innovative strategies. We develop an evolutionary model to show that long-term innovation has to be preserved because, in a competitive context where predator firms exist, long-term innovation cannot be but a dominated strategy.

8.2.1 Basic assumptions

Let us imagine a network of firms, i.e. composed of large and small firms of different age, that decide to join their efforts to develop a major, long-term innovation. This network of long-term innovators is competing with a group of predator firms willing to develop in the same business activity an incremental innovation, less sophisticated and more rapidly available. We assume, as a general principle, that the engagement of investments from firms within the innovation network increases the profitability of long-term innovation, while the engagement of investments from predator firms outside the network decreases the profitability of this long-term innovation. Consequently, a 'critical mass' of firms in the network has to be reached for long-term innovation to be profitable in the long run.

8.2.2 The model

This situation can be formalized on the basis of a two-dimensional linear evolutionary game in which firms come from two strategically distinct populations.[4] Population 1 is composed of network firms, engaged in a long-term innovation, and characterized by a pay-off matrix A. Population 2 is composed of predator firms outside the network, and characterized by a pay-off matrix B. Each population has two alternative actions: to invest (pure strategy 1) or not to invest (pure strategy 2).

The game is specified by the pay-off matrices $A = (a_{hk})$ and $B = (b_{hk})$, where $a_{hk} \in R$ (pay-off to population 1 player) and $b_{hk} \in R$ (pay-off to population 2 player) when population 1 player uses pure strategy h and population 2 player uses pure strategy k.

$$A = \begin{vmatrix} a_{11} & a_{12} \\ a_{21} & a_{22} \end{vmatrix} \quad B = \begin{vmatrix} b_{11} & b_{12} \\ b_{21} & b_{22} \end{vmatrix}$$

The 'critical mass' principle according to which the greater the number of firms in the innovation network willing to invest, the more profitable long-term innovation will be (alternatively the greater the number of

140 *Powerful Finance and Innovation Trends*

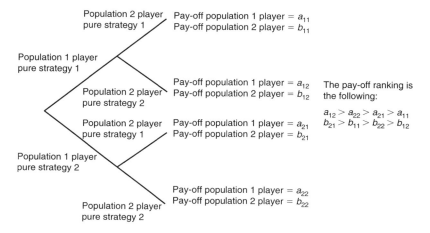

Figure 8.1 Decision tree and pay-off ranking

firms investing in the predator strategy, the lower will be the profitability of long-term innovation) can be expressed by a decision tree (Figure 8.1) and gives the corresponding ranking of returns.

Populations 1 and 2 are considered as large populations of firms. Initially, each population is divided into the fraction $x_1 \in [0, 1]$ (respectively $y_1 \in [0, 1]$) of players in the population currently choosing pure strategy 1, and the fraction $x_2 = 1 - x_1$ ($x_2 \in [0,1]$) (respectively $y_2 = 1 - y_1$, $y_1 \in [0,1]$) of the population choosing pure strategy 2. As the population state (x_1, x_2) (respectively y_1, y_2) changes, so do the pay-offs to the pure strategies. Changes in the population states are governed by the replicator dynamics. Firms are randomly drawn two by two from these populations to play the game (one firm from each player population). If the pay-off to a player in the first (respectively the second) population depends only on the distribution of actions (y_1, $1 - y_1$) in the other population (respectively x_1, $1 - x_1$), the replicator dynamics can be expressed as follows, by a system of time derivatives of the population state (x_1^0, y_1^0) which depends on the pay-off difference between the first and second pure strategies:

$$x_1^0 = [(a_{11} - a_{21})y_1 - (a_{22} - a_{12})y_2]x_1 x_2$$
$$= [(a_{11} - a_{21})y_1 - (a_{22} - a_{12})(1 - y_1)](1 - x_1)x_1$$
$$y_1^0 = [(b_{11} - b_{12})x_1 - (b_{22} - b_{21})x_2]y_1 y_2$$
$$= [(b_{11} - b_{12})x_1 - (b_{22} - b_{21})(1 - x_1)](1 - y_1)y_1$$

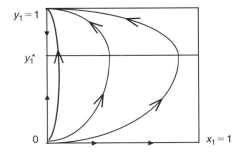

Figure 8.2 Evolution of populations 1 (innovators, x-axis) and 2 (predators, y-axis)

8.2.3 Results

For a large range of parameters, y_1^0 is always increasing, while x_1^0 increases when $y_1 < y_1^*$, and decreases when $y_1 > y_1^*$, with $y_1^* = (a_{11} - a_{22}) / (a_{11} - a_{12} + a_{22} - a_{21})$. In this case, the first action (first column) is dominant for population 2 players and the second action (bottom row) is the best reply by population 1 players. This is characteristic of an iterated dominated strategies game (see Friedman, 1996; Weibull, 1995), where the corner $(x_1, y_1) = (0, 1)$ is the unique Nash equilibrium. This Nash equilibrium is automatically an evolutionary stable equilibrium because it is a solution by iterated elimination of dominated strategies (Figure 8.2). This figure shows that the proportion of population 2 players (predator firms) deciding to engage in investments (labelled y_1) increases monotonically along a large range of solution orbits. The proportion of population 1 players (innovation network firms) deciding to engage in investments (x_1) increases until y_1^* and decreases thereafter.

8.2.4 Comments

Three sets of comments can be derived from the model:

1. The long-term innovative strategy, after being attractive to a growing part of the population of network firms, is adopted less and less. The population of firms willing to undertake such a behaviour decreases to finally equal zero. By contrast, the rival predator strategy becomes more and more prevalent and, in the end, the entire population of firms is effectively engaged in this strategy.
2. The lower the value of y_1^*, the higher the rapidity of extinction of pure strategy 1 in population 1, i.e. the innovative behaviour in the

network. Alternatively, the higher the value of y_1^*, the longer population 1 firms will maintain the innovative behaviour. At the end of the game, predator firms are dominant, but this may result either from a long or a rapid process of competition between predator and long-term innovative firms.

3. The model suggests that external forces have a crucial role to play in whether or not to sustain innovative strategies. If we try to relate the heuristics of this model to what occurs in the real world, we consider that principles of governance that could encourage firms to sustain long-term innovative strategies, up to the point $(x_1, y_1) = (1, 1)$, where innovation is maintained despite predator strategies, are a major form of this kind of external forces.

8.3 The governance of innovative firms

From the conclusions of the model above, we derive that the industry faces two different kinds of evolution. Natural market forces, operating without any external intervention, lead predator firms to dominate the industry all along the ILC, and especially in the later stages. If, on the contrary, investors may want to invest in long-term innovation strategies, then innovative firms have the possibility to survive in the long run. In that case, intervention in favour of innovation must operate beyond the early stages, in order to go against the natural trend of the ILC and, further, ensure the survival of innovative firms in later stages.

This strategy implies encouragement of cooperation between managers and investors that can give support to the development of common learning processes. In formal models, this implies extending the formalization to non-purely adaptive behaviours of firms, i.e. to move from replicator dynamics to learning evolutionary games (Kandori et al., 1993; Samuelson, 1997).[5] In practical terms, for firms' governance, learning processes mean that both managers and investors are jointly committed to develop an entrepreneurial behaviour favouring a long-term perspective, knowledge creation and coordination for innovation, and implying the acceptance of a greater confrontation of uncertainty and higher risk taking. In that perspective, corporate governance is dedicated to the coordination of learning processes, meaning that corporate governance and the governance of knowledge are two facets of the development and coherence of the firm (Penrose, 1959).

This mode of firms' governance has already been analysed by referring to the notion of 'corporate entrepreneurship' (Krafft and Ravix, 2006; see also Foss and Christensen, 2001; Klein and Klein, 2002, for related approaches). Corporate entrepreneurship is a hybrid form mixing the cooperation and control modes described earlier in Section 8.2. Managers, by defining and selecting innovative processes, and investors, by determining the money that is invested to sustain these processes, both take part in the creation and governance of new knowledge by the firm. Cooperation must exist between managers and investors, since they collectively contribute to corporate development and coherence. Only in a second step does control occur: the investor reacts to the innovative choice by validating or refuting the innovative conjecture.

8.3.1 Governance based on corporate entrepreneurship: cooperation for the creation and coordination of new knowledge

Long-term innovation affects corporate coherence, since it involves important reconfigurations of resources and competencies over time (Chatterjee and Wernerfelt, 1991; Foss, 1993; Teece and Pisano, 1994; Teece et al., 1997; Piscitello, 2004). In order to preserve corporate coherence in an innovation context, two sets of problems have to be solved (Foss and Christensen, 2001). First, the problem of knowledge creation, i.e. how knowledge arises from new combinations, and from the discovery of new complementarities between existing stocks of knowledge and experimentation of new learning processes. Second, the problem of coordination of knowledge dispersal, that goes together with the specialization of tasks in large corporations, and involves efforts in constructing shared cognitive patterns.

The role of the manager is crucial in solving these two sets of problems. First because managerial control has a basic facet of stimulating the entrepreneurial activity of the different stakeholders within the firm, but also among the network of innovation partners, that favours knowledge creation. Second because managerial control also involves command, management information systems, corporate routines and corporate cultures, that can act as knowledge coordination. In order to maintain corporate coherence, the manager has a key role to play in the achievement of a critical mass of stakeholders (at the level of the firm and also at the level of the network) playing the long-term innovative strategy, while refraining from the temptation of predator behaviours in the meantime.

The investor is also highly involved in the process. Investors have to develop new competences and experience in the evaluation of long-term innovative companies, since usual market criteria essentially refer to tangible assets and require long-term track records that are often neither applicable nor available in highly innovative contexts. In that perspective, the valuation of intangibles by investors becomes a real issue: valuation is the outcome of a process of coordination of different elements of knowledge related to the perceived ability of the firm to create new technological and market opportunities; valuation is also a key element in sustaining some innovative projects (and not others) that shape the evolution of the industry.

Corporate entrepreneurship means that managers and investors are mostly intertwined *ex ante* in the process of solving corporate coherence problems in the modern cognitive firm. Each actor is endowed with a different piece of knowledge that has to be recombined in a process of collective learning oriented towards corporate development. The manager brings his own competences to the development of learning processes by creating diversity, exploring new opportunities, and providing continuity in innovation. The investor also contributes to the development of learning processes by providing the manager with his own skills and experiences on the financial feasibility of external restructurings (M&As, cooperations), or internal strategies (compensation plans, reporting activities, information systems). Cooperation between managers and investors favours the processes of creation and coordination of new knowledge that are engaged in a long-term innovative context. But the investor also has to control *ex post* the impact of innovative choices implemented in the company.

8.3.2 Governance based on corporate entrepreneurship: control to validate or refute the innovative conjecture

If initially long-term innovation greatly disturbs corporate coherence, one should expect over time that corporate coherence is improved. Here thus, control must be operated to guarantee that coherence is restored or at least evolves towards a reasonable level.

The manager has to provide the investor with regular information (documents, reports, etc.) explaining whether the innovative strategy impacts on corporate coherence and how. If after a sufficiently long time span, the innovative strategy generates new knowledge but is insufficiently coordinated, or if the critical mass of efforts is not obtained and that stakeholders massively turn to adopt the predator strategy, then the manager and the investor can jointly infer that erroneous decisions were taken during the innovation process. The investor checks *ex post* that the manager has implemented the productive and organizational decisions (cooperation agreements or M&As within the innovation network) dedicated to sustain the critical mass of efforts, and further the long-term innovative strategy. Contrary to shareholder value maximization principles, the problem of the investor is not to limit the discretionary power of the manager but, rather, to control this power. In particular, the investor has to control the manager's trustworthiness, and eventually his propensity to 'empire building', in reference to his ability or inability to restore corporate coherence after the engagement of a long-term innovative strategy. When innovation is concerned, thus, the issue is not to impose drastic changes in strategies based on the belief that what the investor (respectively the manager) knows is always right. Rather, the issue is to control the development of innovation and the progressive restoration of corporate coherence.

8.3.3 Summing up

Corporate governance is thus constructed step by step with an *ex ante* process of collective learning, and an *ex post* process of control, in which each actor embodies a piece of diversified and specialized knowledge related to their respective domains and fields of experience, that has to be recombined and used to stimulate corporate development. Table 8.2

Table 8.2 The governance of innovative firms

	The governance of innovative firms
Type of firm	Innovative firms, independent of their age and size
Type of manager	Mix of the professional manager and the entrepreneur
Role of manager	Discover new technological and market opportunities
	Run the company in the interest of all stakeholders
Type of investor	Long-term investors
Role of investor	Add value to the company through active participation
	Assess the value of the company on the basis of economic and financial criteria
	Have a long-term orientation, even if risky
Nature of governance	Corporate entrepreneurship: cooperation, assistance and control
Structure of governance	Different types of committees: scientific, R&D and end-users, audit, compensation systems

provides more details on the governance of innovative firms based on corporate entrepreneurship.

The principle of governance concerns here innovative firms, independent of age and size. The manager has the double role of acting as a professional manager and as an entrepreneur, which means that he has to run the company in the interests of all the stakeholders that contribute to the value of the firm, and also to discover new technological and market opportunities by the active involvement of all stakeholders in learning processes. The investor has to develop a long-term orientation, and is deemed to assess the value of the company on the basis of economic and financial criteria. Cooperation and assistance between the manager and the investor must dominate, on the basis of specific structures of governance favouring common learning processes. One could for instance think of scientific, R&D or end-users committees as structures of governance of this kind. In the meantime, control also has to occur on the basis of more traditional structures of governance, such as audit and compensation systems.

Notes

1. Economics of innovation also provides a clear assessment that innovation networks are composed of large and small firms in interaction (Antonelli, 2003; Saviotti, 1996). Theoretical considerations and empirical observations also suggest that, in many industries, large firms are still important drivers of innovation, either via the development of R&D capacities or via the provision of complementary assets (Teece, 1986; Chesbrough, 2003).
2. Readers not familiar with this body of literature should eventually also refer to exhaustive surveys (Malerba and Orsenigo, 1996; Dosi and Malerba, 2002; Krafft, 2006).
3. The asset owner has the residual right to decide how to use the asset in cases where the contract is silent on the occurrence of some event affecting this use.
4. This model is derived from Foss (1994) and Krafft and Ravix (2005). In the published version, our results are derived using a numerical example. In the present version, more general results are obtained, using a range of parameters.
5. These formal developments are beyond the scope of the current chapter.

References

Aghion, P. and P. Bolton, 'An Incomplete Contracts Approach to Financial Contracting', *Review of Economic Studies*, 59(3) (1992), 473–94.

Antonelli, C., *The Economics of Innovation, New Technology and Structural Change* London: (Routledge, 2003).

Audretsch, D., 'Innovation, Growth and Survival', *International Journal of Industrial Organization*, 13(4) (1995), 441–58.

Audretsch, D. and E. Lehmann, 'Entrepreneurial Access and Absorption of Knowledge Spillovers: Strategic Board and Managerial Composition for Competitive Advantage', *Journal of Small Business Management*, 44(2) (2006), 155–66.

Becht, M., T. Jenkinson and C. Mayer, 'Corporate Governance: an Assessment', *Oxford Review of Economic Policy*, 21(2) (2005), 155–63.

Chatterjee, S. and B. Wernerfelt, 'The Link between Resources and Type of Diversification: Theory and Evidence', *Strategic Management Journal*, 12(1) (1991), 33–48.

Chesbrough, H., *Open Innovation: the New Imperative for Creating and Profiting from Technology* (Cambridge, Mass.: Harvard Business School Press, 2003).

Dosi, L. and F. Malerba (eds), 'Special Issue: Interpreting Industrial Dynamics 20 Years after Nelson and Winter's Evolutionary Theory of Economic Change', *Industrial and Corporate Change*, 11(4) (2002), 3–202.

Foss, N., 'Theories of the Firm: Competence and Strategy Perspectives', *Journal of Evolutionary Economics*, 3(2) (1993), 127–44.

Foss N., 'Cooperation is Competition: George Richardson on Coordination and Interfirm Relations', *British Review of Economic Issues*, 16(40) (1994), 25–51.

Foss, N. and J. Christensen, 'A Market Process Approach to Corporate Coherence', *Managerial and Decision Economics*, 22(4–5) (2001), 213–26.

Fransman, M., *Telecoms in the Internet Age: From Boom to Bust to...* (Oxford: Oxford University Press, 2000).

Friedman, D., 'Equilibrium in Evolutionary Games: Some Experimental Results', *Economic Journal*, 106(434) (1996), 1–25.

Geroski, P., 'What do we Know about Entry?', *International Journal of Industrial Organization*, 13(4) (1995), 421–40.

Gompers, P., 'Optimal Investment, Monitoring, and the Staging of Venture Capital', *Journal of Finance*, 50(5) (1995), 1461–90.

Gompers, P. 'Grandstanding in the Venture Capital Industry', *Journal of Financial Economics*, 42(1) (1996), 133–56.

Gompers, P. and J. Lerner, 'The Venture Capital Revolution', *Journal of Economic Perspectives*, 15(2) (2001), 145–68.

Gort, M. and S. Klepper, 'Time Paths in the Diffusion of Product Innovations', *Economic Journal*, 92(367) (1982), 630–53.

Grossman, S. and O. Hart, 'The Costs and Benefits of Ownership: a Theory of Vertical and Lateral Integration', *Journal of Political Economy*, 94(4) (1986), 691–720.

Hart, O., *Firms, Contracts and Financial Structure* (Oxford: Oxford University Press, 1995a).

Hart, O., 'Corporate Governance: Some Theory and Implications', *Economic Journal*, 105(430) (1995b), 678–89.

Hart, O. and J. Moore, 'Incomplete Contracts and Renegotiation', *Econometrica*, 56(4) (1988), 755–86.

Jensen, M. and W. Meckling, 'Theory of the Firm: Managerial Behaviour Agency Costs and Ownership Structure', *Journal of Financial Economics*, 3(4) (1976), 305–60.

Kandori, M., G.J. Mailath and R. Rob, 'Learning, Mutation and Long-Run Equilibria in Games', *Econometrica*, 61(1) (1993), 29–56.

Kaplan, S. and P. Stroemberg, 'Financial Contracting Theory Meets the Real World: an Empirical Analysis of Venture Capital Contracts', *Review of Economic Studies*, 70(243) (2003), 281–315.

Kaplan, S. and P. Stroemberg, 'Characteristics, Contracts, and Actions: Evidence from Venture Capitalist Analyses', *Journal of Finance*, 59(5) (2004), 2177–210.
Klein, P. and S. Klein, "Do Entrepreneurs Make Predictable Mistakes? Evidence from Corporate Divestitures', in N.J. Foss and P.G. Klein (eds), *Entrepreneurship and the Firm: Austrian Perspectives on Economic Organization* (Cheltenham: Edward Elgar, 2002).
Klepper, S., 'Industry Life Cycles', *Industrial and Corporate Change*, 6(1) (1997), 119–43.
Krafft, J., 'What do We Know about Industrial Dynamics? Introduction to the Special Issue', *Revue de l'OFCE*, June.
Krafft, J. and J.L. Ravix, 'The Governance of Innovative Firms: an Evolutionary Perspective', *Economics of Innovation and New Technology*, 14(3) (2005), 125–48.
Krafft, J. and J.L. Ravix, 'Corporate Governance and the Governance of Knowledge: Rethinking the Relationship in Terms of Corporate Coherence', *Economics of Innovation and New Technology*, 17(1), (2008).
Lazonick, W. and M. O'Sullivan (eds), *Corporate Governance and Sustainable Prosperity* (New York: Palgrave Macmillan, 2002).
Malerba, F. and L. Orsenigo, 'The Dynamics of Evolution of Industry', *Industrial and Corporate Change*, 5(1), 51–87.
Penrose, E., *The Theory of the Growth of the Firm* (London: Basil Blackwell, 1959).
Piscitello, L., 'Corporate Diversification, Coherence and Economic Performance', *Industrial and Corporate Change*, 13(5) (2004), 757–87.
Samuelson, L., *Evolutionary Games and Equilibrium Selection* (Cambridge, Mass. and London: MIT Press, 1997).
Saviotti, P.P., *Technological Evolution, Variety and the Economy* (Cheltenham: Edward Elgar, 1996).
Schleifer, A. and R. Vishny, 'A Survey on Corporate Governance', *Journal of Finance*, 52(2) (1997), 737–83.
Teece, D. 'Profiting from Innovation', *Research Policy*, 15(6) (1986), 285–305.
Teece, D. and G. Pisano, 'The Dynamic Capabilities of Firms: an Introduction', *Industrial and Corporate Change*, 3(3) (1994), 537–56.
Teece, D., G. Pisano and A. Shuen, 1997, 'Dynamic Capabilities and Strategic Management', *Strategic Management Journal*, 18(7) (1997), 509–33.
Weibull, J., *Evolutionary Game Theory* (Cambridge, Mass.: MIT Press, 1995).
Williamson, O., *The Economic Institutions of Capitalism* (New York: Free Press, 1985).
Williamson, O., 'The Institutions of Governance', *American Economic Review*, 88(2) (1998), 75–9.
Williamson, O., 'The New Institutional Economics: Taking Stock, Looking Ahead', *Journal of Economic Literature*, 38(3) (2000), 595–614.

9
Economic Change and the Organization of Industry: Is the Entrepreneur the Missing Link?

Edouard Barreiro and Joël Thomas Ravix

9.1 Introduction

In the economic literature the entrepreneur is usually associated with innovation and change. Actually we can distinguish several kinds of entrepreneurs who are essentially gathered into two traditions. The first, the Austrian approach, describes the entrepreneur as the agent of novelty and change. The second, called the 'Anglo Saxon tradition' (Witt, 1999, 2005), considers the firm as the central unit of analysis. As a consequence the entrepreneur is only a virtual agent who exists but who has no real functions in the work of the economic system. Although we cannot imagine an entrepreneur without a firm, 'economic theorizing has not taken much interest in the relation between entrepreneurship and the firm as an organizational form' (Witt, 1999, p. 99).

However, it is possible to distinguish a third way: the Marshallian, which uses a different approach of knowledge to connect innovation and economic development to the firm, through the role of the entrepreneur. As a result, the entrepreneur organizes production but is also the engine of economic change. This framework is less used mainly for two reasons. First, there is the systematic opposition from the neoclassical Marshall, who developed in the *Principles of Economics* the theory of demand and supply and the partial equilibrium, and the evolutionist and empirical approach of *Industry and Trade*. If the primary approach could be accepted, unless it remains a simplified view, the second is more questionable. Indeed, Marshall is therefore at the origin of a tradition in which concepts and analyses were developed by others. In fact, his writings hold few metaphoric references to evolutionist principles. Secondly, the economic literature rarely goes beyond the presentation of organization as a fourth factor of production or as a phenomenon that operates at three distinct levels: the firm, the industry and the nation. That kind of interpretation masks the fact that for Marshall 'organization aids knowledge' and 'knowledge is our most powerful engine of production' (Marshall, 1920, p. 115). Moreover, these three levels of organization are not conceptually different; it is just the opposite, they are all relative to the organization of production which has two dimensions: one specific to the firm, 'the internal' organization, one specific to the industry, 'the external organization'. If the former enables us to explain the origin of increasing returns the latter lets us foresee the future potential economies. The problem becomes how to articulate these two dimensions.

Marshall considers two solutions to overcome this difficulty. The first, the most famous, rests on the concept of the representative firm and the distinction between internal and external economies. This attempt leads, however, to a dead end because Marshall does not manage 'to link' his static approach to value and his dynamic conception of industrial development (Thomas, 1991; Quéré and Ravix, 1998). The second solution, fuzzier, appears in Marshall's analyses of the role of the 'manufacturer' or of the 'entrepreneur' (Loasby, 1986). Indeed, in his *Principles*, he insists on the fact that the entrepreneur 'who makes goods not to meet special orders but for the general market, must, in his first role as merchant and organizer of production, have a thorough knowledge of *things* in his own trade' (Marshall, 1920, p. 248). In this extended knowledge, Marshall includes the capacity of 'forecasting the broad movements of production and consumption, of seeing where there is an opportunity for supplying a new commodity' and to improve 'the plan of producing an old commodity' (ibid.). In integrating the economic function of knowledge in the role of the entrepreneur, Marshall opens an original analytical perspective likely to explain the conceptual link between innovation and industrial organization. The interest of this new perspective is to show that innovation is not independent from the action of producing and that to analyse the former and its effects, it is necessary to be able to grasp the latter, i.e. to theorize on the economic function of the entrepreneur.

In this chapter, we propose to show that the field of research opened up by Marshall was partly explored by Frank H. Knight (1921) with his concept of the organic entrepreneur. But the entrepreneur function was diluted by Ronald H. Coase (1937) to explain the nature of the firm.

This result, which can be called 'the paradox of the entrepreneur', also appears in a different way in the Austrian approach and more specifically in I. Kirzner's work. We will see in a third section that unlike the new institutional economy, the Austrians tackle the question of the action of the entrepreneur. Lastly, in a fourth section, we will show how E. Penrose avoids this paradox and converges to the Marshallian approach in managing simultaneously to treat the firm as an institution and to give to the action of the entrepreneur a precise theoretical status.

9.2 Frank H. Knight and the organic entrepreneur

For Knight, 'the uncertainty problem in economics is the forward-looking character of the economic process itself' (Knight, 1921, p. 237). That is what he explains thanks to the fact that 'the production of goods requires time' (ibid.). More precisely, uncertainty has two origins: first, decisions have to be made before the launching of the production process and no one knows if it will be successful. Secondly, the demand is in reality future needs, which are also hard to predict. As a result: 'the producer, then, must *estimate* (1) the future demand which he is striving to satisfy and (2) the future results of his operations in attempting to satisfy that demand' (ibid., p. 238).

Knight is convinced that 'under the enterprise system, a special social class, the business men, direct economic activity; they are in the strict sense the producers, while the great mass of the population merely furnish them with productive services, placing their persons and their property at the disposal of this class' (ibid., p. 271). But the function of the entrepreneur is not limited to the organization and the direction of production, 'the entrepreneurs also guarantee to those who furnish productive services a fixed remuneration' (ibid.). This last role is a necessary condition to guarantee that the entrepreneur will be able to manage production. In fact, 'with human nature as we know it, it would be impracticable or very unusual for one man to guarantee to another a definite result of the latter's actions without being given power to direct his work. And on the other hand the second party would not place himself under the direction of the first without such a guaranty' (ibid., p. 270).

Anyway, the role of the entrepreneur cannot be defined only through the specificity of profits compared to any other kind of payment. It is more related to the question of the place of the entrepreneur in the mutation of the general conditions of economic activity. Knight has this approach. Indeed, 'Knight's philosophy is to recall the distinction he constantly makes between the mechanical and organic (biological) frameworks. Mechanistic thinking views human behaviour and institutions as static, machine-like entities, whereas organicistic thinking invokes notions such as change and process' (Langlois and Cosgel, 1993, p. 458). This distinction, essential for Knight, enables us to understand his thought on the role of knowledge and on 'the relation between knowledge and behaviour' (Knight, 1921, p. 197). He insists specifically on the fact that 'if all changes were to take place in accordance with invariable and universally known laws, they could be foreseen for an indefinite period in advance of their occurrence, and would not upset the perfect apportionment of product values among the contributing agencies, and profit (or loss) would not arise' (ibid., p. 198). Hence, for Knight, 'it is our imperfect knowledge of the future, a consequence of change, not change as such, which is crucial for the understanding of our problem' (ibid.). The main idea is not that change occurs or could occur in the economic process but that we are unable to predict future events. However, if our ignorance about the future is radical, we would be in a similar situation than if all the events, past, present and future, are known. In both situations nobody can predict anything and there would be no entrepreneurs and no profits. On the contrary, when some random events occur with 'sufficient regularity to be practically predictable in large measure' (ibid.), the principle of anticipation becomes meaningful. Hence emerge 'the justification and the necessity for separating (...) the effects of change from the effects of ignorance of the future' (ibid.). This conceptual distinction permits us to understand why some agents dare to anticipate although the future is unknown. Indeed, 'even though the business man could not know in advance the results of individual ventures, he could operate and base his competitive offers upon accurate foreknowledge of the future if quantitative knowledge of the probability of every possible outcome can be had' (ibid., pp. 198–9).

Be that as it may, while the entrepreneur manages uncertainty, he is not for Knight a speculator. Speculation is not specific to the entrepreneur's function and it is done by others such as the insurance agent or the fund holder. On the other hand, the entrepreneur is the only one who assumes the risk induced by the production temporality. This activity is not limited to technical coordination, in a mechanical world. Indeed, 'with uncertainty entirely absent, every individual being in possession of perfect knowledge of the situation, there would be no occasion for anything of the nature of responsible management or control of productive activity' (ibid., p. 267). In such a world, 'the flow of materials and productive services through productive processes to consumer would be entirely automatic' (ibid.). The set-up of such an automated system would be 'the result of a long process of experimentation, worked out by trial-and-error methods alone' (ibid.). This mechanical world does not exclude the presence of 'managers, superintendents, etc., for the purpose of coordinating the activities of individuals' (ibid., p. 267), but the last are 'labourers merely, performing a purely routine function, without responsibility of any sort' (ibid., p. 268).

The introduction of uncertainty completely changes the situation and 'the primary problem or function is deciding what to do and how to do it' (ibid.). According to these conditions, production cannot be entrusted to anybody who coordinates the production factors only in a mechanical way. The function of the producers is increasingly complex, for two reasons:

> In the first place, goods are produced for a market, on the basis of an entirely impersonal prediction of wants, not for the satisfaction of the wants of the producers themselves. The producer takes the responsibility of forecasting the consumers' wants. In the second place, the work of forecasting and at the same time a large part of the technological direction and control of production are still further concentrated upon a very narrow class of the producers. (Ibid.)

According to Knight, these reasons justify the existence of a new actor: the entrepreneur. This one is essential because 'when uncertainty is present and the task of deciding what to do and how to do it takes the ascendancy over that of execution, the internal organization of the productive groups is no longer a matter of indifference or mechanical detail' (ibid.). As a consequence, the organization of production could no longer be done in a mechanical and routinized way; a total mutation of the producer function occurs. This metamorphosis is set up as a biological process since 'centralization of this deciding and controlling function is imperative, a process of "cephalisation", such as has taken place in the evolution of organic life, is inevitable, and for the same reasons as the case of biological evolution' (ibid., pp. 268–9).

This process leads to an 'organic entrepreneur' (Quéré and Ravix, 1997), who has, essentially, two ways to buffer uncertainty. On the one hand, 'consolidation', which relies on the 'law of large numbers' and, on the other, 'specialization', which consists in the selection of the most capable in the management of uncertainty. Following Knight, 'consolidation and specialization are intimately connected' (ibid., p. 240) because they are related to a characteristic of human nature: 'Men differ in their capacity by perception and inference to *form correct judgments* as to the future course of events in the environment' (ibid., p. 241). However, as these capacities can be gained with experience, Knight admits that 'knowledge is more a matter of learning than of the exercise of absolute judgment' (ibid., p. 243). But 'learning requires time, and in time the situation dealt with, as well as the learner, undergo change' (ibid.). As a result the entrepreneur originates in an endogenous and cumulative process. This process can be explained first because 'the specialisation of uncertainty-bearing in the hands of entrepreneurs involves also a further consolidation'; and secondly it is 'closely connected with changes in technological methods which (a) increase the time length of the production process and correspondingly increase the uncertainty involved, and (b) form producers into large groups working together in a single establishment or productive enterprise and hence necessitates concentration of control' (ibid., p. 245).

9.3 The rise of the firm as an institution and the eviction of the entrepreneur

Coase generates the firm from the market. This approach constrains him to give up the prospect developed by Knight and to compare the entrepreneur to a simple coordinator, with no specific role. The aim of Coase is to show 'that a definition of a firm may be obtained which is not only realistic in that it corresponds to what is meant by a firm in the real world, but is tractable by two of the most powerful instruments of economic analysis developed by Marshall, the idea of the margin and that of substitution, together giving the idea of substitution at the margin' (Coase, 1937, p. 386). 'The Marshall' used by Coase is obviously the one of the Book V of the *Principles* who explains how the allocation of resources is carried out by the price system. It is thus not astonishing that Coase considers that 'Marshall introduces organization as a fourth factor of production' (ibid., p. 388).

By adopting this point of view, Coase partly changes the debate of his time from the role of the entrepreneur to the problem of compatibility between two alternative modes of allocation of resources: the firm and the market. He wishes 'to bridge what appears to be a gap in economic theory between the assumption (made for some purposes) that resources are allocated by means of the price mechanism and the assumption (made for other purposes) that this allocation is dependent on the entrepreneur-co-ordinator' (ibid., p. 389). The real aim of Coase is to explain 'why a firm emerges at all in a specialized exchange economy' (ibid., p. 390). As a consequence, he develops an analysis to compare the allocative efficiency of two institutions: the firm and the market. These are in competition as they both have the same functions but different mechanisms. Indeed, in the market allocation is done through the price system whereas in the firm it is under the authority of the entrepreneur. We can underline that in this perspective the entrepreneur has the same function as the market.

The choice of the best modality of coordination is based on the assumption that the information is imperfect. In this respect the use of the price mechanism has a cost as the research of the relevant prices or negotiating. Nevertheless, in 'forming an organization and allowing some authority (an entrepreneur) to direct resources, certain marketing costs are saved'. For Coase, the uncertainty is exclusively about the course of the transactions. This point of view is far from the analyses of Knight (1921), and if Coase recognizes that 'it seems improbable that a firm would emerge without the existence of uncertainty', he makes clear that 'those, for instance, Professor Knight, "who make the mode of payment" the distinguishing mark of the firm (...) appear to be introducing a point which is irrelevant to the problem we are considering' (Coase, 1937, p. 392).

The problem is the size of the firm, or of its boundaries, i.e. the share of the function of coordination in between the firm and the market, which according to Coase, relies on the 'diminishing returns to management' (ibid., p. 395). This static problem can be solved with the adoption of a sharing rule. For Coase, the organization of a transaction in-house also has a cost. As a consequence, 'a firm will tend to expand until the costs of organizing an extra transaction within the firm become equal to the costs of carrying out the same transaction by means of an exchange on the open market or the costs of organizing in another firm' (ibid., p. 395). The size of the firm is then a function of the marginal cost of an operation. Beyond the equilibrium position, 'the costs of organizing within the firm will be equal either to the costs of organizing in another firm or to the costs involved in leaving the transaction to be "organized" by the price mechanism' (ibid., p. 404).

The firm appears from the market and is designed to allocate resources at a cost lower or equal to the market one. This function seems to be specific enough for Coase to justify the role of the entrepreneur. Indeed, he claims that his analysis 'clarified the relationship between initiative or enterprise and management' (ibid., p. 405), functions that are both, according to him, combined by the entrepreneur. However, 'initiative means forecasting and operates through the price mechanism by the making of new contracts' and 'management proper merely reacts to price changes, rearranging the factors of production under its control' (ibid.). As a result the functions of the entrepreneur are in reality governed by the price system. He is not different from the other agents of the economic system and like them he behaves according to the signs of the market.

The eviction of the entrepreneur is clearer as Coase's analysis is carried out in the static framework of equilibrium. It is really hard to understand what can be the initiative function of the entrepreneur in such a construction. This contradiction appears in *the nature of the firm* when he explains that 'business men will be constantly experimenting, controlling more or less, and in this way, equilibrium will be maintained. This gives the position of equilibrium for static analysis' (ibid., p. 404). As a consequence, there is not a specific implication of the entrepreneur in the functioning of the economy; and nothing makes it possible to specify his role.

Although Coase evokes the idea that the 'dynamic factors' have an important influence on the costs of organizing an activity through the firm and the market, his analysis remains basically static for two reasons. Firstly, he clarifies neither the nature of these dynamic factors nor their incidence; secondly, he does not differentiate the action of coordination and that of allocation. On the other hand, if we retain the idea suggested by M. Casson that 'coordination is a dynamic concept, as opposed to allocation, which is a static one', the entrepreneur can appear since 'the concept of coordination captures the fact that the entrepreneur is an agent of change: he is not concerned merely with the perpetuation of the existing allocation of resources, but with improving upon it' (Casson, 1982, p. 24). However, such a perspective implies giving up the traditional interpretation of the idea of coordination but also leaving the static framework of the Coasian analysis.

The problem of Coase is not the same that Knight proposed solving about 20 years before. This is clear in the divergent points of view that these two authors have about uncertainty. Indeed, Coase retains a narrow view of uncertainty, taking the form of imperfect information on the whole states of the world. In spite of its static nature, it is sufficient to justify that the transactions organized within the firm or on the market are more or less expensive and to explain the emergence of the firm as an institution that coordinates as the market. For Knight, uncertainty has another dimension. It is basically related to the temporal character of production and to the fact that the economic process itself changes. In his own way Knight extends the Marshallian vision and avoids the paradox of the entrepreneur met by Coase, but he does not really approach the problem of the institutional organization of production.

9.4 Market process and the dilution of the entrepreneur's action

The Austrian perspective can be approached as a second junction because it does not follow Coase (1937) but Hayek (1937). As a result, instead of being interested in the firm as an organization, they point out the role of knowledge in the action of the entrepreneur on economic activity. 'Entrepreneurship is about change. It is about how the organization of economic activity extends and reshapes itself. The theme of novelty and change is especially clear in Schumpeter, for whom entrepreneurship is the carrying out of new combinations, and in Kirzner, for whom entrepreneurship is the perception of new frameworks of means and ends' (Langlois, 2005, p. 3). Anyway, we can distinguish these two approaches. Schumpeter (1934) insists on the entrepreneur as an innovator, though Kirzner emphasizes the role of arbitration of the entrepreneur (1973). They both criticize the neoclassical theory of competition, but on different bases. If Schumpeter questions the logical solidity and the relevance of the theory of perfect competition, Kirzner in contrast seeks to provide a better comprehension of the forces leading to the determination of the equilibrium.

Nevertheless, it is possible to admit that 'in the competitive market process, the Schumpeterian and Kirznerian entrepreneurs may complement each other – the one creating change, the other responding to it' (McNulty, 1987, p. 537). Kirzner has the same point of view and points out that 'Schumpeter's entrepreneur acts to *disturb* an existing equilibrium situation. Entrepreneurial activity disrupts the continuing circular flow. The entrepreneur is pictured as *initiating* change and as generating new opportunities.' Although 'each burst of entrepreneurial innovation leads eventually to a new equilibrium situation, the entrepreneur is presented as a disequilibrating rather than an equilibrating force' (Kirzner, 1973, p. 72). By contrast, says Kirzner, 'my own treatment of the entrepreneur emphasizes the equilibrating aspect of his role' (ibid., p. 73). The adoption of this interpretation enables him to stress the voluntarism of the entrepreneur in the animation of the market process.

Indeed, the analysis of Kirzner rests on a precise distinction between the concepts of market equilibrium and market process. When we reach the equilibrium all the agents' decisions are mutually compatible and perfectly coordinated because all of them have complete information concerning the decisions of the other participants. However, for Kirzner, in such a situation the activity of the entrepreneur does not have a *raison d'être*. On the contrary, when coordination is not complete, the entrepreneur plays an essential role. Thanks to his 'alertness' to seize profit opportunities, the entrepreneur drives the market process. His action contributes to supporting a better coordination of the agent's plans to allow equilibrium of the market.

This concept of 'alertness in action' enables us to explain why 'far from being numbed by the inescapable uncertainty of our world, men *act upon their judgments of* what opportunities have been left unexploited by others' (Kirzner, 1973, p. 87). But the use of this concept of 'alertness' to characterize the entrepreneur leads Kirzner to 'emphasize the capture of pure entrepreneurial profit as essentially reducible to the exploitation of arbitration opportunities' (Kirzner, 1982, p. 141). Indeed, he thinks that there is a formal similarity between buying and selling on various markets today or at different dates.

Ced Kirzner starts from the notions of knowledge and discovery which are, according to him, specific of the Austrian approach, in order to design the concept of the arbitrating entrepreneur. Indeed, 'this approach (a) sees equilibration as a systematic process in which market participants acquire more and more accurate and complete *mutual knowledge* of potential demand and supply attitudes, and (b) sees the driving force behind this systematic process in (...) *entrepreneurial discovery*' (Kirzner, 1997, p. 62). The characteristic of this approach is to use the idea of 'sheer ignorance' and not the hypothesis of imperfect information, because 'sheer ignorance differs from imperfect information in that the discovery which reduces sheer ignorance is necessarily accompanied by the element of *surprise*' (ibid.). However, this element of surprise could not be compared to a simple cost of the research or the production of missing information. Under these conditions, Kirzner can conceive the entrepreneurial discovery as a process which is 'gradually but systematically pushing back the boundaries of sheer ignorance, in this way increasing mutual awareness among market participants and thus, in turn, driving prices, output and input quantities and qualities, toward the values consistent with equilibrium' (ibid.). The engine of this process is the permanent existence of profit opportunities which, while leading the market towards equilibrium, prevents it ever being reached. Indeed, 'except in the never-attained state of complete equilibrium, each market is characterized by opportunities for pure entrepreneurial profit. These opportunities are created by earlier entrepreneurial errors which have resulted in shortages, surplus, misallocated resources' (ibid., p. 70). In this framework, the action of the entrepreneur favours the convergence of the market towards equilibrium since 'the daring, alert entrepreneur discovers these earlier errors, buys where prices are "too low" and sells where prices are "too high"' (ibid.).

However, this behaviour of arbitrage does not allow such a process to succeed insofar as the actions of correction of the entrepreneur open new opportunities which, once seized, can transform former actions into errors. The market process evolves continuously for two reasons. On the one hand, 'that continual change in tastes, resource availabilities, and known technological possibilities always prevent this equilibrative process from proceeding anywhere near to completion', and on the other, 'that entrepreneurial boldness and imagination can lead to pure entrepreneurial losses as well as to pure profit. Mistaken actions by entrepreneurs mean that they have misread the market, possibly pushing price and output constellations in directions not equilibrative' (ibid., p. 72).

Kirzner's emphasis on the entrepreneur's action joins L. von Mises who points out 'in any real and living economy every actor is always an entrepreneur and a speculator' (Mises, 1949, Part 4, Ch. XIV). Under these conditions, it becomes difficult to define with precision the boundaries of the entrepreneur's functions. Indeed, for him 'every action is embedded in the flux of time and therefore involves speculation'. As a result the function of the entrepreneur is totally diluted since 'the capitalists, the landowners, and the labourers are by necessity speculators. So is the consumer in providing for anticipated needs' (ibid.). Consequently, to recognize an entrepreneur among buyers and sellers it is necessary to give him a specific psychological profile. This is the *raison d'être* of the concept of 'natural alertness' which comes to justify that the contractor is not an economic agent like the others. However, this is not sufficient to differentiate Kirzner's entrepreneur from that of Mises.

More generally, the idea according to which the entrepreneur perceives before others the profit opportunities generated by the existence of uncertainty, raises an important problem perfectly identified by G.B. Richardson (1960) about the theory of equilibrium. Even within the framework of perfect information, 'the existence of such a general profit potential cannot automatically be assumed to create particular profit opportunities for individual entrepreneurs'. Indeed, Richardson makes clear, 'before any particular entrepreneur is prepared to invest in the production of commodity, he will have to be assured that the volume of supply planned by competing producers, who are also aware of the opportunity, will not be so large as to overstock the market, thus converting the expectation of profit into the realization of loss'. But as Richardson points out, 'how, in a perfect market, where all producers are free to move in response to the profit opportunity, is that assurance to be afforded him? And yet without this assurance, entrepreneurs would not invest, and supply would not be expanded; a general profit potential, which is known to all, and equally exploitable by all, is, for this reason, available to no one in particular' (Richardson, 1960, p. 14). This paradox, which is due to the fact that competition prevents agents from communicating between themselves, can be perfectly extended to a situation of uncertainty. In this environment, the entrepreneur cannot know if the opportunity he perceives is identified by others, but he also does not know if this opportunity of profit is mistaken. The paradox is reinforced since uncertainty, judicious to justify the role of the entrepreneur, prevents him from acting.

Two solutions can be considered to overcome this difficulty. The first is developed by Richardson himself. He goes further, with the idea that entrepreneurs devise their investments. By extension, the hypothesis makes it possible to solve the question of the boundaries of the firm (Richardson, 1972), but does not bring new elements about the status of the entrepreneur in the firm. The second solution, which is not contradictory with the first, is developed by Edith T. Penrose in *The Theory of the Growth of the Firm* (1959). This work enables us to explain not only the institutional dimension of the firm but also the behaviour of the entrepreneur.

9.5 The entrepreneurial firm as a solution to the paradox of the entrepreneur

Penrose's analysis of the growth of the firm is better known than that of the entrepreneur. However, as we will see, the first cannot be conceived without the second. Indeed, she points out that the firm is 'the basic unit for the organization of production' and because it has this particular function it is 'a complex institution, impinging on economic and social life in many directions, comprising numerous and diverse activities, making a large variety of significant decisions, influenced by miscellaneous and unpredictable human whims' (Penrose, 1959, p. 9). By tackling the question of the firm from the perspective of production and not of exchange, Penrose distinguishes herself from Coase (Ravix, 1999) and shows that this has a double dimension: on the one hand, it is an institution which has the function of carrying out production. On the other, as this function implies making decisions, the firm also has a behavioural dimension which concerns the action of the entrepreneur. This is the double dimension which makes it possible to define the concept of the entrepreneurial firm, defined in Penrose's analyses.

The institutional dimension is important to justify the specificity of the firm compared to the market: 'the essential difference between economic activity inside the firm and economic activity in the "market" is that the former is carried on within an administrative organization, while the latter is not' (Penrose, 1959, p. 15). However, Penrose goes beyond the concept of direction, characteristic of the administrative organization. If this dimension allows establishment of a distinction between organization and market, it is not specific to the firm since other forms of socio-economic organizations do the same. That is why Penrose insists on the fact that 'a firm is more than an administrative unit; it is a collection of productive resources the disposal of which between different uses and over time is determined by administrative decisions' (ibid., p. 24).

Penrose emphasizes the problem of the organization of production and establishes an additional distinction between productive resources and the services rendered by these resources. Thus, 'strictly speaking, it is never *resources* themselves that are the "input" in the production process, but only the *services* that the resources can render' (ibid., p. 25). For Penrose, 'the important distinction between resources and services is not their relative durability; rather it lies in the fact that resources consist of a bundle of potential services and can, for the most part, be defined independently of their use, while services cannot be so defined, the very word "service" implying a function, an activity' (ibid.).

Penrose starts from this distinction to dismiss the concept of production factor and to replace it by 'productive opportunity'. This concept does not indicate the entirety of the average materials available in a firm at a given time, but corresponds to the second dimension of the firm distinguished by Penrose. Indeed, the productive opportunity of the firm, by gathering 'all of the productive possibilities that its "entrepreneurs" see and can take advantage of' (ibid., p. 31), makes it possible to characterize the behavioural dimension of the firm, its action of undertaking, which completes its institutional dimension. Penrose links this second dimension with the concept of entrepreneurship which, even if it is a 'slippery' concept, is closely associated with 'the temperament or personal qualities of individuals' (ibid., p. 33), and remains nevertheless an essential element to understand the process of the growth of the firms. The concept of 'productive opportunity' can also be treated as 'a psychological predisposition on the part of individuals to take a chance in the hope of gain, and, in particular, to commit effort and resources to speculative activity' (ibid., p. 33). However, this psychological predisposition has nothing to do with the quality of the anticipation or of the calculations of the entrepreneur. It simply corresponds to 'the decision to make some calculations' (ibid.).

Penrose considers that 'the "expectations" of a firm – the way in which it interprets its "environment" – are as much a function of the internal resources and operations of the firm as of the personal qualities of the entrepreneur' (ibid., p. 41). This report enables her to establish, within her concept of 'productive opportunity', a distinction between the potential 'objective' productive opportunity of the firm, expressing what the firm is able to do or its competences, and its 'subjective' productive opportunity, corresponding to what the firm 'thinks it can accomplish' (ibid.). If the objective productive opportunity is linked to the internal resources and the activity of the firm, the subjective productive opportunity indicates how the firm interprets its environment.

However, Penrose highlights that ' "expectations" and not "objective facts" are the immediate determinants of a firm's behaviour' (ibid., p. 41). This idea is an extension of K. Boulding's work (1956), according to which there are not objective facts but only 'images' subjectively created from the interpretation of information coming from the environment.

> In other words, rather than beginning with the objective environment of the firm, and the information that this environment generates – in the form, for example, of market prices, market demands, the activities of competitors, etc. – Penrose starts with the mental world of the planners who are situated within the context of their own firm and its specific productive services. (Fransman, 1994, p. 743)

Indeed, the only objective phenomena are irreversible past events that the firm cannot any longer handle, while the firm's expectations rest on possibilities not yet realized. As a result, 'firms not only alter the environmental conditions necessary for the success of their actions, but, even more important, they know that they can alter them and that the environment is not independent of their own activities' (Penrose, 1959, p. 42).

With this framework Penrose avoids the problem raised by Richardson and provides a satisfactory explanation of entrepreneurial behaviour. Indeed, in her approach uncertainty does not block the entrepreneur's action. The latter is not a simple arbiter who speculates while adapting to market opportunities. He is a real entrepreneur who tries to transform his environment to take advantage of it. This analysis is based on a concept of risk and uncertainty different from that of Knight. As the environment of the entrepreneur cannot provide objective data, it results from it that ' "uncertainty" refers to the entrepreneur's confidence in his estimates or expectations; "risk", on the other hand, refers to the possible outcomes of action, specifically to the loss that might be incurred if a given action is taken' (ibid., p. 56). This opposition arises from the idea that decision-making is not related to the various possibilities which could be realized, but to the way the entrepreneur interprets these possibilities. Thus the risk is not linked to the probability of realization of a random event, but to the material capacities of the firm to assume it. Under these conditions, the possibilities of action of the entrepreneur are not directly related to the risks inherent to his environment, but depend on his resources and on his competences. Thus we are able to understand why the action of the entrepreneur necessarily takes place in the firm. Indeed, as Langlois points out, 'the firm exists because of entrepreneurship' (2005, p. 1). As a consequence, entrepreneurial action implies the creation and development of a company.

Because one admits Penrose's idea of 'the essentially subjective nature of demand from the point of view of the firm' (ibid., p. 80), it becomes possible to consider that the entrepreneur does not take 'demand as "given", but rather as something he ought to be able to do something about' (ibid.). In other words, the market is not external to the firm, but a normal extension of its activity. It is similar to the commercial deals set up by the firm since 'the "demand" with which an entrepreneur is concerned when he makes his production plans is nothing more nor less than his own ideas about what he can sell at various prices with varying degrees of selling effort' (ibid., p. 81). More precisely, according to Penrose, to answer this demand – that she perceives in a subjective way – the entrepreneur builds 'areas of specialization' which represent both the internal and external organization of the firm. This concept of areas of specialization includes two different but narrowly complementary elements: on the one hand, the 'production base', which gathers all the means and technical skills mobilized by the company to produce. On the other hand, the 'market area' corresponds to 'each group of customers which the firm hopes to influence by the same sales programme' (ibid., p. 110). The concept of areas of specialization comes to complete that of productive opportunity. These two concepts are thus respectively merely the concretization of the institutional dimension and the behavioural dimension of the firm.

9.6 Conclusion

The abandonment of the Marshallian perspective, by both the neo-institutional approach and by the Austrian one, leads to the eviction or the dilution of the role of the entrepreneur. This result is very paradoxical concerning the Austrian approach as its ambition is to put the action of the entrepreneur at the core of the market process. A similar attempt can be found in Knight's work, but it should not be confused with the Austrian one. Indeed, these authors not only justify the economic role of the entrepreneur by his psychological profile, they give him the mission to guarantee a fixed income to the economic agents who do not like to take risks. However, this function is missing in the analyses of Mises and Kirzner because the first, while adhering to the catalectic, considers economic activity as the only activity of exchange, and the second focuses only on the market process. As a consequence, both pay no attention to the specificity of the productive dimension, which Knight does. However, this function is not very helpful in justifying the role of the entrepreneur since it appears as the result and not as the cause of his action. Indeed, because the entrepreneur sets up production, he is led to pay a fixed income to employees and to capitalists and not the reverse. Interpreted like this, the criticism of Coase with regard to Knight is relevant: the fact of paying incomes to certain categories of agents is not sufficient to specify the function of the entrepreneur.

Coase puts aside the problem of the function of the entrepreneur to rise the issue of the nature of the firm. Thus he manages to define the simplest institutional divisions of the activity of coordination: between the firm and the market. Although this dichotomy occurs in an uncertain environment, it remains static and does not permit us to treat the issue of the action of the entrepreneur. In contrast, Penrose emphasizes the productive function of firms; consequently she manages to explain that these act because 'the environment is not something "out there", fixed and immutable, but can itself be manipulated by the firm to serve its own purposes' (Penrose, 1995, p. xiii). With this very specific approach to the environment, Penrose is able to show that the possibilities of action of the firm are closely related to its internal and external organization, i.e. to its productive opportunities and its areas of specialization which are, respectively, the result of its accumulated experience. That is the reason why she claims that 'one of the primary assumptions of the theory of the growth of the firm is that "history matters"' (ibid.). This accumulated experience, which comes from the past activities of firms, appears in the 'changes in knowledge acquired and changes in the ability to use knowledge'. Indeed, to change its area of specialization, each firm has to acquire new knowledge and create new competences and 'change in knowledge acquired and changes in the ability to use knowledge' (Penrose, 1959, pp. 52–3). To modify its area of specialization each firm has to acquire new knowledge to create new competences and 'this increase in knowledge not only causes the productive opportunity of a firm to change in ways unrelated to changes in the environment, but also contributes to the "uniqueness" of the opportunity of each individual firm' (ibid.). This process, by explaining how knowledge acts on production and how the organization helps knowledge, may be used as an analytical basis to understand the institutional organization of production in the extension of Marshall's work.

As a conclusion, the firm is not a static element of the economic system. Its evolution leads to a perpetual change in the internal and external organization of production. But the firm does not behave by itself, it is the entrepreneur who drives this engine. He is at the origin of each decision because he is the only one which can interpret and determine how to act on the environment. As a consequence he is the driving belt which transmits the change to the industry and to the whole economic system.

References

Boulding, K. E., *The Image* (Ann Arbor, Mich.: University of Michigan Press, 1956).
Casson, M., *The Entrepreneur* (Oxford: Martin Robertson, 1982).
Coase, R. H., 'The Nature of the Firm', *Economica*, 4 (November 1937).
Fransman, M., 'Information, Knowledge, Vision and Theories of the Firm', *Industrial and Corporate Change*, 3 (5) (1994), 713–57.
Hayek, F. A., 'Economics and Knowledge' (1937), in F. A. Hayek, *Individualism and Economic Order* (London: Routledge and Kegan Paul, 1948).
Kirzner, I. M., *Competition and Entrepreneurship* (Chicago: University of Chicago Press, 1973).
Kirzner, I. M., 'Uncertainty, Discovery, and Human Action: a Study of the Entrepreneurial Profile in the Misesian System', in I. M. Kirzner (ed.), *Method, Process, and Austrian Economics, Essays in Honor of Ludwig von Mises* (Lexington, Mass.: Lexington Books, 1982).
Kirzner, I. M., 'Entrepreneurial Discovery and the Competitive Market Process: an Austrian Approach', *Journal of Economic Literature*, 35 (1) (1997), 60–85.
Knight, F. H., *Risk, Uncertainty and Profit* (1921) (London: Reprints of the London School of Economics and Political Science, 1948).
Langlois, R. N., 'The Entrepreneurial Theory of the Firm and the Theory of the Entrepreneurial Firm', Working papers 2005–27 (Hartford: University of Connecticut, Department of Economics, 2005).
Langlois, R. N. and M. M. Cosgel, 'Frank Knight on Risk, Uncertainty and the Firm: a New Interpretation', *Economic Inquiry*, XXXI (3) (1993), 456–65.

Loasby, B. J., 'Marshall's Economics of Progress', *Journal of Economic Studies*, 13 (5) (1986), 16–26.
Marshall, A., *Principles of Economics* (1920) (London: Macmillan, 1979, 8th edn).
McNulty, P. J., 'Competition: Austrian Conceptions', in J. Eatwell, M. Milgate and P. Newman (eds), *The New Palgrave: a Dictionary of Economics*, vol. 1 (London: Macmillan, 1987).
Mises, L. von, *Human Action. A Treatise on Economics* (Irvington-on-Hudson, NY: The Foundation for Economic Education, 1949).
Penrose, E. T., *The Theory of the Growth of the Firm* (Oxford: Basil Blackwell, 1959).
Penrose, E. T., *The Theory of the Growth of the Firm* (Oxford: Oxford University Press, 1995, 3rd edn).
Quéré, M. and J.-L. Ravix, 'Le chercheur entrepreneur dans la dynamique des relations science-industrie', in B. Guilhon, P. Huard, M. Orillard and J.-B. Zimmermann (eds), *Economie de la connaissance et organisations* (Paris: L'Harmattan, 1997).
Quéré, M. and J.-T. Ravix, 'Alfred Marshall and Territorial Organization of Industry', in M. Bellet and C. L'Harmet (eds), *Industry, Space and Competition* (Cheltenham: Edward Elgar, 1998).
Ravix, J.-T., 'Edith T. Penrose and Ronald H. Coase on the Nature of the Firm and the Nature of Industry', in C. Pitelis (ed.), *The Growth of the Firm. The Legacy of Edith Penrose* (Oxford: Oxford University Press, 1999).
Richardson, G. B., *Information and Investment* (Oxford: Clarendon Press, 1960).
Richardson, G. B., 'The Organisation of Industry', *The Economic Journal*, 82 (327) (1972), 883–96.
Schumpeter, J. A., *The Theory of Economic Development* (Cambridge, Mass.: Harvard University Press, 1934).
Thomas, B., 'Alfred Marshall on Economic Biology', *Review of Political Economy*, 3 (1) (1991), 1–14.
Witt, U., 'Do Entrepreneurs Need Firms? A Contribution to a Missing Chapter in Austrian Economics', *The Review of Austrian Economics*, 11 (1–2) (1999), 99–109.
Witt, U, 'Firms as Realizations of Entrepreneurial Visions', Papers on Economics and Evolution 2005–10 (Berlin: Max Planck Institute of Economics, Evolutionary Economics Group, 2005).

10
Innovation and Profitability in the Computer Industry

Christian Genthon

10.1 Introduction

The purpose of this chapter is to explore the relationship between innovation and profitability by analysing the computer industry over an extended period of time (20 years). The aim is to identify the determinants of innovation, taking this particular industry as an example. To do so, our first section will analyse the industry's industrial dynamics, while the second will present conclusive research regarding the determinants of R&D. We will then empirically test the predictions generated by these two analytical frameworks. We find that the intensity of R&D expenditure bears no relation to profitability. But the main findings of this chapter are that the determinants of innovation can differ within the same industry, at different times.

The essay is divided into three parts. Section 10.2 presents an analysis of the computer industry's dynamics over a long period. Section 10.3 introduces the methods and results that have already been produced on the topic of the relationship between innovation and profitability. Section 10.4 is an attempt to empirically validate our analysis of the computer industry and relations between innovation and profits.

10.2 Industrial dynamics of the computer industry

A sector's industrial organization results from the interaction between a certain number of variables. These variables relate primarily to four factors: the technological system dimension (knowledge base, appropriability, etc.) and the technical base (scale economies, for example), the type and degree of competition, the strategies implemented and finally the performance of the players. Our hypothesis is that, in every industry, over a given period of time, a specific dynamic exists between the variables mentioned above. This industrial organization has the properties of a structure and possesses certain permanence. The dynamic stability of the system is ensured by the consistency of the relationship between the variables. But innovation, whether it comes from inside or outside the industry, can bring into question the existing industrial organization, which in turn can lead to a period of crisis and transition followed by the setting up of a new industrial organization. It should be noted that innovations are generally absorbed by the existing industrial organization without notable modifications. In other words, sector-based dynamics must be seen on two levels: so-called minor changes, whose influences on the sector have little significance, and so-called major changes, which change the course of industry.

In the computer industry,[1] a break occurred in the 1990s prompted by the invasion of standardization in an industry that had rejected it for decades. Indeed, the industry's organization between the 1950s and the end of the 1980s was based precisely on the *incompatibility* between competing products/systems: each competitor used one or several *proprietary* operating systems, which ensured that its customers would remain loyal. The situation was one of monopolistic competition with native differentiation of products. The dynamics of innovation, created in the United States around spin-offs and venture capital, never brought this organization into question. Innovation was even developed on the basis of these principles: a new application or product was based on a specific operating system and was, therefore, naturally proprietary. It can be noted that entry barriers decrease proportionally as we move away from large systems (mainframes), because user lock-in depends on the amount of investment required in specific software. Indeed, the form of industrial organization that was based on the incompatibility of operating systems placed innovation at the heart of its dynamic. Thanks to their generic R&D capabilities, the sector's larger firms were able to catch up with the innovators and share the market the most successful of them had opened. A continuous innovation/catch-up cycle was established as the use of computers became widespread.

Let us take the example of the PDP8 minicomputer invented by DEC (which later became Digital) in 1965. Not only did this innovation allow it to become number two in the computer sector 20 years later, it also allowed new firms to develop in this segment, such as Hewlett-Packard and Data General. The major players copied it, especially IBM with its System 7, which was succeeded by the S36, the S38 and, finally, the

AS400, which is at the beginning of the twenty-first century the number one minicomputer family in the world. The same phenomenon occurred a few years later on the workstation market. Workstations were pioneered by firms such as Ridge Computer, Apollo, Intergraph, Silicon Graphics and Sun Microsystems at the beginning of the 1980s. When the market became more mature, the leading manufacturers of scientific minicomputers, such as Hewlett-Packard and Digital, decided to enter it. IBM arrived a little late in the day, but succeeded in catching up. In 1989, the market shares of the three leaders (Sun, Hewlett-Packard and Digital) were 18, 17 and 15 per cent respectively. Five years later, the three leaders were Sun, IBM and Hewlett Packard with 20, 18 and 17 per cent of market share respectively (sources: *Datamation* and companies' annual reports). The figure did not change a great deal after that point. The microcomputer's initial development became part of this framework: Apple, Tandy and Commodore, the three leaders between 1977 and 1981, offered machines that were mutually incompatible because they used proprietary operating systems.

R&D is central in this type of industrial organization. Innovation enables new firms to invent new products and create new markets. R&D capabilities enable established firms to set up catch-up strategies. For this rationale to be coherent, pricing policy had to cater for significant R&D spending. This form of industrial organization lasted until IBM altered the rules of the game at the beginning of the 1980s, by introducing a product, the PC, with an operating system it did not control. This break was to launch the era of standardization, which would cause a structural change in the sector.

The new form of industrial organization, based on standards and open systems, drove the industry towards mass production and the reduction of diversity. Standards first emerged at PC level (processors and operating systems) and then diffused to other subsectors such as workstations, minicomputers and servers. The standardization process affected not only the PC's central processing unit, but also its peripherals (floppy disk, hard disk, monitor, printers, etc.). Consequently, the vertical integration that had characterized the sector disappeared almost entirely and spare parts (mass memory, peripherals) began to be produced by specialist suppliers who enjoyed scale economies. Price competition took the place of competition based on differentiation. The production phase of the industrial process (assembly) was relocated to South East Asia (the so-called Four Dragons) and then to China. Product differentiation, i.e. customers made captive by proprietary operating systems, also vanished from the greater part of the market (microcomputers, servers, workstations). Diversity or the number of different operating systems and processors used, diminished considerably. The consequence of this change is that oligopolistic competition is today based mainly on marketing and, to a minor extent, technological innovation and R&D expenditure. Therefore, R&D no longer holds the central role it used to have. Another consequence of the change in industrial organization was the structural crisis the industry suffered for many years. Incumbent firms experience difficulties with an often endless adjustment process and new firms are not yet large enough to shape the way the industry is organized.

Is this interpretative framework congruent with the empirical data and, if so, what can we surmise regarding the relationship between innovation and profit? More broadly, what are the determinants of innovation? Much work has been carried out on innovation in the computer industry but few people have worked from statistical series.[2] On the other hand, most studies on the relationship between innovation and profits have been cross-sectional, and have only been able to focus limited attention on the sectoral dimension. What we propose is an analysis of a long statistical series relating to a single industry, the computer industry. But before we do so, we must present the topic. We will restrict our study to the theme of the relationship between innovation, profitability and sectoral industrial organization.

10.3 Innovation, profits and industrial organization

The three key terms of our study can be studied in pairs. We should specify that innovation will be measured according to R&D expenditure. The R&D expenditure/sales ratio will be called R&D intensity.

10.3.1 Innovation and profit

Innovation provides a competitive advantage because it allows a monopolistic or quasi-monopolistic position to be maintained until competitors are able to imitate it. It brings about greater profits, but this advantage is invariably temporary. But one can imagine that some companies play the innovation game systematically. On the other hand, the cost of imitation is, as a general rule, lower than that of innovation. Research has shown that a firm's R&D activities can benefit its competitors (spillover) (Jaffe, 1986). The trade-off between innovation and imitation depends on cost and timing, and we can assume that the relationship between innovation and profits depends on the sectoral pattern. Recent work on this theme generally uses cross-sectional autoregressive models of the following type:

Performance = a + b Performance (−1) + c Innovation Variables
+ d Control Variables (firms, sectors)

For instance, Geroski et al. (1993) show that a particular innovation increases profits (the operating income/turnover ratio) by around 6 per cent. They employ the SPRU database, which identifies commercially successful innovations in Great Britain. Leiponen (2000) finds a positive link between profit and innovation. He uses the innovation survey performed by the Finnish government over the course of the 1990s. He draws a distinction between process innovation and product innovation. His results show that product innovation has a negative effect on profits. Cefis and Ciccarelli (2005) distinguish between innovators and non-innovators using patent information (267 English companies from 1988 to 1992). They believe that innovation has a positive effect on a firm's profitability. But they do note that 'Results are not robust across the different estimation methods' (p. 60). As in the case of the Geroski et al. sample, innovative companies are larger than non-innovative ones. Hula (1988), who bases his examination on the first *Business Week R&D Survey* (year 1977), fails to establish any relationship between R&D spending and profit. Lastly, cross-sectional studies do not appear to bring about a consensus regarding the existence of a relationship between innovation and profits.

10.3.2 Innovation and industrial organization

Our hypothesis is that the status of R&D in the computer industry changed between 1982 and 2001. In the 1980s, R&D was a necessity for industrial firms. The growth of the business went hand in hand with an increased emphasis on R&D, regardless of the firm's economic situation. In the 1990s, R&D budgets did not increase in proportion to the growth of firms and, to some extent, we support the hypothesis that R&D has now become an adjustment variable used when a company faces difficulties (decrease in turnover and profitability, unbalanced financial structure). As we have already mentioned, the industrial organization of the computer industry in the 1970s and 1980s was based on the incompatibility of the products/systems of different manufacturers. This was an era of high product differentiation, of monopolistic competition. The dominant companies imposed competition through R&D, with the average rate of R&D expenditure rising between 1982 and 1990. The new form of industrial organization that appeared in the 1990s was based on standards that were, and still are, not the property of the manufacturers.[3] R&D became less strategic because it barely allowed firms to differentiate themselves. R&D was no longer at the heart of the industrial organization of the computer industry: its status had to evolve along with the change in industrial organization.

10.3.3 Profit and industrial organization

Farrell et al. (1998), in their seminal article on the vertical organization of industry, analyse the differences between two forms of competition, the type they call 'system competition' (or closed competition) and the type they call 'component competition' (or open competition). They refer explicitly to the computer industry. Their model (a Bertrand one) shows that system competition induces higher profits than component competition. They hypothesize that firms do in fact choose the form of competition that allows higher profits. In the case of the computer industry, the two organizational models have coexisted for many years. Newcomers cannot choose the closed system and incumbent firms switch over to the open system with great difficulty. It should be noted that the authors do not present an empirical evaluation.

On the subject of modular organization versus vertical integration, the body of literature has steadily increased (cf. Langlois and Robertson, 1992; Baldwin and Clark, 1997, to quote just two articles). But one can note that few empirical works have attempted to measure the phenomenon. To our knowledge, only Schilling and Steensma (2001) have put forward both a theoretical view and an empirical test. They identify two forces driving the phenomenon of flexible organization: heterogeneous inputs and demand that induces the need for complementary competencies. They add the following catalyst variables: availability of standards, technical change and competitive intensity. The results of their empirical work show that heterogeneity and standards have a significant effect on modularity, measured in terms of contract manufacturing, alternative work arrangement and alliances. Their analysis is cross-sectional and static. It does not allow us to identify changes within an industry, if indeed they exist.

We put forward the hypothesis that the closed model was dominant in the computer industry in the 1980s, only to be succeeded by the open model in the 1990s–2000s. One could presume that vertically integrated firms (closed system) were more profitable in the 1980s, and less so in the 1990s–2000s. To test our hypothesis, we will not use the variables employed by Schilling and Steensma. Chiefly, this is because variables that are suitable for cross-sectional analyses may not be totally appropriate for single-industry studies. Vertical integration is very difficult to measure, which is why Schilling and Steensma use a bundle of three variables. In the computer industry, vertical integration has historically been linked to proprietary systems. For this particular industry, we can construct a variable representing the degree of vertical integration: the volume of proprietary systems a firm sells as a proportion of its total sales.

10.4 Empirical analysis

10.4.1 Database and variables

Each year, the industry's top 60 firms (SIC 3571 in the American nomenclature) are ranked according to their computer-related turnover. The database is comprised of the subset of firms that make more than 50 per cent of their turnover in the computer industry.[4] The database is therefore unbalanced and is considered to be representative of the industry each year. The database contains figures for the following variables for the 20 years between 1982 and 2001: total turnover, R&D expenditure, net income, operating income, shareholders' equity, total assets, (mainframe + mini) turnover, maintenance turnover and proprietary systems turnover. We have constructed a seven-point Likert scale to identify the degree of proprietorship of each firm's computer systems each year. The scale's construction is based on the following quantitative data: (mainframe turnover + mini turnover)/total turnover, maintenance turnover/total turnover (*Datamation* and annual reports) and qualitative data: proprietary turnover/total turnover (press and annual reports). In addition, three periods are explored: 1982–89 (first industrial organization), 1994–2001 (second industrial organization) and the entire database (1982–2001). It is clear that the transitional period does not exactly correspond to the years 1990–93. Every incumbent firm was impacted differently at different times. We therefore tried other subdivisions (for example 1983–89 and 1993–2000) and we obtained robust results when using these different time periods (cf. Appendix for more information on the database).

Many studies use a time lag of one year to allow for the impact of R&D, as is the case in Cohen et al. (1987). Others believe that a time lag of two to three years is required because of the long-term effects of R&D. Fanchon (2003) used a data envelopment analysis to select appropriate variables for the measurement of efficiency in the computer industry. He

Table 10.1 Descriptive statistics

%	NI/TO 1982–2001	NI/TO 1982–1989	NI/TO 1994–2001
Mean	1.5	3.7	0.7
Standard error	8.4	7.7	7.4
Median	3.2	4.9	1.9
Max	21.4	21.4	11.8
Min	−38.0	−31.8	−33.4
N	520	213	183
%	RDI 1982–2001	RDI 1982–1990	RDI 1994–2001
Mean	7.9	9.1	5.7
Standard error	4.2	3.3	4.0
Median	7.9	9.2	5.8
Max	20.3	17.5	18.2
Min	0.16	1.6	0.16
N	520	213	183

notes that 'past R&D expenditure seems to have little effect on current sales' (p. 180). We tried using a three-year mean, but this did not change the results and so we kept the one-year lag for R&D.

The descriptive statistics relating to the net income/turnover (NI/TO) and R&D expenditure/turnover (RDI) ratios are provided in Table 10.1.

10.4.2 Innovation and profit

We specify the following model (Tables 10.2 and 10.3):

$$\text{Performance} = a + b \text{ Performance\_1} + c \text{ IRD\_1} + d \text{ GRTO} + \text{dummy variables}$$

We use three performance ratios: net income/turnover (NI/TO), operating income/turnover (OI/TO) and shareholders' equity/total assets (SE/TA). IRD_1 is the R&D intensity of the previous year. GRTO represents year-on-year turnover growth. The financial results of firms in a dynamic industry such as computers can be dependent on the growth of their sales.

The dummy variables Europe and Japan are also included. Some European companies received state grants which allowed them to survive

Table 10.2 Relationship between profitability and R&D intensity, 1982–89

	NI/TO	OI/TO	SE/TA
Constant	−0.02	0.008	0.059*
	(−1.48)	(0.69)	(2.29)
(Dependent variable)_1	0.600*	0.791*	0.885*
	(9.01)	(16.21)	(22.76)
IRD_1	0.002	−0.0003	−0.002
	(1.15)	(−0.25)	(−0.14)
GRTO	0.062*	0.025	−0.02
	(3.83)	(1.78)	(−0.91)
Europe	−1.320	−0.009	−0.0002
	(−0.11)	(−0.81)	(0.01)
Japan	0.005	0.005	−0.004
	(0.27)	(0.30)	(−0.17)
R^2	0.391	0.667	0.868
N	190	162	154
Method	OLS	OLS	OLS

( ): t value; *: $p < 0.01$.

Table 10.3 Relationship between profitability and R&D intensity, 1994–2001

	NI/TO	OI/TO	SE/TA
Constant	−0.005	−0.007	0.257*
	(−0.48)	(−0.47)	(7.96)
(Dependent variable)_1	0.246*	0.138	0.346*
	(3.39)	(1.92)	(6.12)
IRD_1	0.0004	0.000	0.006
	(0.34)	(0.35)	(2.11)
GRTO	0.107*	0.160*	0.041
	(4.54)	(6.17)	(1.18)
Europe	−0.026	0.000	−0.183*
	(−1.50)	(0.01)	(−3.70)
Japan	−0.005	0.02	−0.166*
	(−0.29)	(0.64)	(−3.06)
R^2	0.233	0.260	0.632
N	176	170	169
Method	OLS	GLS	GLS

( ): t value (z value when GLS); *: $p < 0.01$.

even with poor results. As regards Japanese companies, it was once thought that they behave in a different way from common American capitalism (cf. Hundley et al., 1996). What is more, these companies are conglomerates. Duysters and Hagedoorn (2001) have demonstrated that the computer industry's internationalization has not yet led to a process of convergence between the strategies and structures of the different companies in the Triad (US, Europe and Japan).

We used the Breusch–Pagan test to assess the heteroscedasticity problem. The results of these tests (see the Appendix) allowed us to use the ordinary least squares (OLS) method for the years 1982–89. However, for the years 1994–2001 it was necessary to use the random effects generalized least squares (GLS) model, for the variables OI/TO and SE/TA.

Some of the results of the regressions contradict what is commonly represented. Firstly, there is no relationship between R&D expenses and profitability. This result is obtained whatever the profit rate we use. The growth rate of turnover is always significant, except in the case of the SE/TA ratio for the years 1994–2001. In the case of the dummy variables Europe and Japan, they are not significant, except in the case of the SE/TA ratio for the years 1994–2001. There is clear interaction between these last two results.

10.4.3 Innovation, profit and industrial organization

The independent variable we select to represent the two different periods of the computer industry is the proprietary-system intensity. We specify two models, one relating to the relationship between profits and industrial organization, the other concerning the relationship between innovation and industrial organization.

The first of these autoregressive models is

$$\text{Performance} = a + b \text{ Performance}\_1 + c \text{ PROP} + d \text{ TOGR}$$
$$+ \text{ dummy variables (Europe, Japan)}$$

We use the same performance ratios as previously: net income/turnover (NI/TO), operating income/turnover (OI/TO) and shareholders' equity/total assets (SE/TA) (see Tables 10.4 and 10.5).

The dummy variables Europe and Japan are not presented. As expected, they are not significant, except for the years 1994–2001 in the case of the SE/TA ratio (with a negative sign). The results of the regressions show that the variable PROP is significant in the first period, but not in the second, whatever the profit rate being used. The Breusch–Pagan test gives the same results as previously.

The second autoregressive model is

$$\text{RDI} = a + b \text{ RDI}\_1 + c \text{ PROP} + d \text{ MS} + \text{dummy variables}$$

Table 10.4 Relationship between profits and industrial organization, 1982–89

1982–89	NI/TO	OI/TO	SE/TA
Constant	−0.038*	−0.011	0.029
	(−2.89)	(−0.90)	(1.08)
(Dependent variable)_1	0.550*	0.746*	0.880*
	(8.25)	(14.37)	(23.60)
PROP	0.006*	0.004	0.005
	(2.81)	(1.75)	(1.68)
TOGR	0.075*	0.033	−0.009
	(4.61)	(2.26)	(−0.42)
R^2	0.416	0.673	0.871
N	199	162	158
Method	OLS	OLS	OLS

( ): t value; *: $p < 0.01$.

Table 10.5 Relationship between profits and industrial organization, 1994–2001

1994–2001	NI/TO	OI/TO	SE/TA
Constant	−0.013	−0.028	0.278*
	(−1.03)	(−0.16)	(7.52)
(Dependent variable)_1	0.254*	0.132	0.364*
	(3.56)	(1.87)	(6.42)
PROP	0.003	−0.000	0.002
	(0.98)	(−0.02)	(0.40)
TOGR	0.112*	0.159*	0.034
	(4.64)	(6.05)	(0.94)
R^2	0.233	0.257	0.626
N	177	170	169
Method	OLS	GLS	GLS

( ): t value (z value when GLS); *: $p < 0.01$.

The market share (MS) variable has been added to the model. It allows us to control for the size of the companies and to test a traditional theme in literature on innovation: the relationship between innovation and company size (on this subject, cf. Cohen et al., 1987; Cohen and Levin, 1989; Cohen and Klepper, 1996, among others) (see Table 10.6).

R&D intensity is linked to the degree to which proprietary systems are used. R&D intensity is not linked to the size of the firms (all the regressions are non-significant to the usual 5 per cent level). Therefore, the size of the firm does not seem to be a determinant of R&D intensity.

Table 10.6 Determinants of R&D

	1982–1989	1994–2001	1982–2001
Constant	0.315	0.538	−0.173
	(1.09)	(1.79)	(−3.051)
RDI_1	0.842*	0.626*	0.893*
	(21.79)	(14.23)	(43.360)
PROP	0.238*	0.449*	0.204*
	(4.05)	(4.50)	(5.854)
MS	−0.011	−0.078	−0.010
	(−0.69)	(−1.89)	(−0.940)
Europe	0.054	−1.03	−0.113
	(0.21)	(−1.98)	(−0.881)
Japan	−0.301	0.330	0.051
	(−0.72)	(0.46)	(0.396)
R^2	0.850	0.930	0.999
N	195	179	501
Method	OLS	GLS	OLS

( ): t value (z value when GLS); *: $p < 0.01$.

These results confirm our hypothesis that R&D was an imperative during the 1980s: the PROP coefficient is twice as high in the second period (0.449) as in the first (0.238).

10.5 Conclusions

Let us sum up our findings:

- There is no relationship between R&D and profits.
- Profits are linked to growth.
- There have been two main periods in the computer industry. This is strongly confirmed by the statistical analysis: OLS for the first period and GLS for the second (random effects have to be taken in account).
- Profits are linked to proprietary systems in the first period, but not in the second.
- R&D is linked to proprietary systems.

Based on the characteristics of technological opportunities and appropriability, Malerba and Orsenigo (1996) introduced the 'technological regime' concept with the aim of identifying groups of industries possessing similar determinants. They demonstrate that there are two fundamental technological regimes, presented in the literature as Schumpeter

Mark I and Schumpeter Mark II. We believe it is possible to go even further and look at the determinants of innovation in detail. Generally, it can be said that innovation processes are *industry-specific*. Indeed, the generic indicators of appropriability and cumulativity combine in the computer industry in a very specific way, in the form of proprietary systems and vertical integration. It is difficult to identify these indicators in cross-sectional studies. Finally, our study reveals more specifically that the determinants of innovation are not only industry-specific but also *historically industry specific*. Indeed, the innovation system (or technological system) of a sector is part of/is an element of/is a dimension of the sector's industrial organization. The innovation system is a subset of the industrial organization of an industry. The innovation system is specific to a particular moment in a given industry's lifetime. This can account for the difficulty in establishing generic laws.

Appendix

A.1 Representativeness of the sample

The sample's representativeness has to be checked from two perspectives:

- That of the sample, that is: computing turnover of the firms in our sample/total computing turnover of the industry's top 60 firms. We obtain a value between 74 and 78 per cent.
- That of the *Business Unit* logic: the computer turnover/total turnover ratio of our sample varies between 85 and 90 per cent depending on the years.

We also need to check the representativeness of the top 60 firms vis-à-vis the entire industry. Global turnover generated by the industry is obviously unknown, but all indications are that the top 60 firms represent a major section of the industry.

A.2 Breusch–Pagan estimators: var $(u) = 0$

Table 10A.1

	NI/TO	OI/TO	SE/TA
Chi2 (1)	0.05	0.19	0.15
Prob > chi2	0.82	0.66	0.70

Table 10A.2

	NI/TO	OI/TO	SE/TA
Chi2 (1)	2.09	7.00	11.04
Prob > chi2	0.15	0.00	0.00

Table 10A.3

	NI/TO	OI/TO	SE/TA
Chi2 (1)	0.37	0.01	0.18
Prob > chi2	0.54	0.94	0.67

Table 10A.4

	NI/TO	OI/TO	SE/TA
Chi2 (1)	1.32	4.93	7.77
Prob > chi2	0.25	0.03	0.00

Table 10A.5

	1982–89	1994–2001	1982–2001
Chi2 (1)	0.08	6.99	0.02
Prob > chi2	0.77	0.01	0.89

Notes

1 Regarding the computer industry, see among others Brock (1975), Stoneman (1976), Flamm (1987, 1988), Fransman (1995), Van den Ende and Kemp (1999), Bresnahan and Malerba (1999), Bresnahan and Greenstein (1999) and Genthon (2004).
2 Momigliano (1983) is an exception.
3 The fact that some standards are proprietary (Intel and, above all, Microsoft) is not relevant to our discussion.
4 To ensure statistical continuity, a firm whose computing/total ratio falls temporarily below 50 per cent remains listed. This only concerns one Japanese firm (NEC).

References

Baldwin, C. and K. Clark, 'Managing in an Age of Modularity', *Harvard Business Review*, 75 (1997), 84–93.

Bresnahan, T. and S. Greenstein, 'Technological Competition and the Structure of the Computer Industry', *The Journal of Industrial Economics*, 67 (1) (1999), 1–40.

Bresnahan, T. and F. Malerba, 'Industrial Dynamics and the Evolution of Firms' and Nations' Competitive Capabilities in the World Computer Industry', in D. Mowery and R. Nelson (eds), *The Sources of Industrial Leadership* (Cambridge: Cambridge University Press, 1999).

Brock, G., *The US Computer Industry: a Study of Market Power* (Cambridge: Ballinger Publishing, 1975).

Cefis, E. and M. Ciccarelli, 'Profit Differentials and Innovation', *Economics of Innovation and New Technology*, 14 (2005), 43–61.

Cohen, W. and S. Klepper, 'A Reprise on Size and R&D', *The Economic Journal*, 106 (1996), 925–51.

Cohen, W. and R. Levin, 'Empirical Studies of Innovation and Market Structure', in R. Schmalensee and R. Willig (eds), *Handbook of Industrial Organisation* (Amsterdam: North-Holland, 1989).

Cohen, W., R. Levin and D. Mowery, 'Firm Size and R&D Intensity: a Re-examination', *Journal of Industrial Economics*, 35 (1987), 543–63.

Datamation, The EDP One Hundred, annual (until 1996).

Duysters, G. and J. Hagedoorn, 'Do Company Strategies and Structure Converge in Global Markets? Evidence from the Computer Industry', *Journal of International Business Studies*, 32 (2) (2001), 347–56.

Fanchon, P., 'Variable Selection for Dynamic Measures of Efficiency in the Computer Industry', *International Advances in Economic Research*, 9 (3) (2003), 175–86.

Farrell, J., H. Monroe and G. Saloner, 'The Vertical Organization of Industry: Systems Competition versus Component Competition', *Journal of Economics & Management Strategy*, 7 (2) (1998), 143–82.

Flamm, K., *Targeting the Computer: Government Support and International Competition* (Washington: The Brookings Institution, 1987).

Flamm, K., *Creating the Computer: Government, Industry and High Technology* (Washington: The Brookings Institution, 1988).

Fransman, M., *Japan's Computer and Communications Industry* (Oxford: Oxford University Press, 1995).

Genthon, C., *Analyse sectorielle: Méthodologie et application aux technologies de l'information* (Paris: L'Harmattan, 2004).

Geroski, P., S. Machin and J. Van Reenen, 'The Profitability of Innovating Firms', *RAND Journal of Economics*, 24 (1993), 198–211.

Hula, D., 'Advertising, New Product Profit Expectations and the Firm's R&D Investment Decisions', *Applied Economics*, 20 (1988), 125–42.

Hundley, G., C. Jacobson and S. Park, 'Effects of Profitability and Liquidity on R&D Intensity: Japanese and U.S. Companies Compared', *Academy of Management Journal*, 39 (1996), 1659–74.

Jaffe, A., 'Technological Opportunity and Spillovers of R&D: Evidence from Firms' Patents, Profits, and Market Value', *American Economic Review*, 76 (1986), 984–1001.

Langlois, R. and P. Robertson, 'Networks and Innovation in a Modular System: Lessons from the Micro-computer and Stereo Component Industries', *Research Policy*, 21 (1992), 297–313.
Leiponen, A., 'Competencies, Innovation and Profitability of Firms', *Economics of Innovation and New Technology*, 9 (2000), 1–24.
Malerba, F. and L. Orsenigo, 'Schumpeterian Patterns of Innovation are Technology-specific', *Research Policy*, 26 (1996), 451–78.
Momigliano, F., 'Determinanti ed effetti della R&D in un'industria ad alta opportunità tecnologica: un'indagine econometrica', *L'industria*, 4 (1) (1983), 61–109.
Schilling, M. and K. Steensma, 'The Use of Modular Organizational Forms: an Industry-level Analysis', *Academy of Management Journal*, 44 (6) (2001), 1149–68.
Stoneman, P., *Technological Diffusion and the Computer Revolution* (Cambridge: Cambridge University Press, 1976).
Van den Ende, J. and R. Kemp, 'Technological Transformations in History: How the Computer Regime Grew out of Existing Computing Regimes', *Research Policy*, 28 (1999), 833–51.

11
Creation and Co-evolution of Strategic Options by Firms: an Entrepreneurial and Managerial Approach of Flexibility and Resource Allocation

Thierry Burger-Helmchen

The success of a firm depends on its capability to create and exploit new projects, routines and technologies. Those projects are competitive opportunities that the firm must recognize, evaluate, and on which it has to be able to apply operating capabilities to take advantage of them. The general management responsible for the firm's strategic direction frequently fails to manage the organization's technological innovation process that creates these opportunities (Adner and Levinthal, 2004). To help managers in their decision-making process in uncertain environments new techniques and theories are developed; one of them is the real option theory. This conceptual decision-making framework is about to become a standard. The formal approach, originating from financial models, dealing with future uncertainty and the opportunities a firm can seize, is appealing to managers.

A major issue, almost ignored in this literature, is the question of the origin of the real options a firm possesses. A suggested answer is the use of the entrepreneur/manager duality to explain the creation of new options. The insights gained from considering real options and entrepreneurship are bidirectional. On the one hand, entrepreneurship, in a resource-based framework, can explain the origin of real options and contribute to a better evaluation of their value. On the other hand, real options can explain the direction a decision maker gives to the development of the new capabilities and resources, as an entrepreneurial activity, by suggesting another use of those resources. Combining entrepreneurship and real options explains the heterogeneity of the firm and its resources collection and capability building. A conceptual model of the option chain (the successive steps of creation, development and use of an option) is presented by Bowman and Hurry (1993).

This conceptual representation highlights the necessity for the firm to change its structure (between entrepreneurial, physical and human capital) in order to be able to explore and create new options and exploit them accurately to survive the competition at the industry level. This work formalizes this conceptual representation into a simulation model.

The next section of the chapter outlines the real option approach and its implication in strategic management for describing exploration and exploitation decisions. The second section presents the model. The following two sections detail the results of the model. The final section wraps up the results.

11.1 Exploration–exploitation representation as an option chain

In the strategic theory of the firm (Teece et al., 1997) the most obvious of the long-term goals of the firm is its survival. Survival is achieved by seizing profitable opportunities when they arise or are encouraged to emerge. In the long run the profitability, survival and growth of a firm do not depend so much on the general efficiency with which it is able to organize production as it does on the ability of the firm to establish one or more wide and relatively impregnable competence 'bases' from which it can adapt and extend its operations in an uncertain, changing and competitive world (Penrose, 1959, p. 137).

An option gives the right but not the obligation to take a specific decision (invest, defer, alter) on an underlying asset, for a predetermined price at, or before, a certain time. For example, a firm can possess a production plant, and choose, depending on customer demand or competition, to build a bigger capacity plant to achieve economies of scale (a growth option) or, to the contrary, to temporarily shut down the plant (option to defer production). The firm has the right, but not the obligation, to change its production capacity. This option, depending on the information at hand at the moment of exercise, allows the firm to catch new revenue flows or to reduce costs.

Figure 11.1 represents the successive developments of the real option chain in the strategic management literature. The following discussion and presentation of the option chain and the introduction of the entrepreneur/manager rely on this figure.

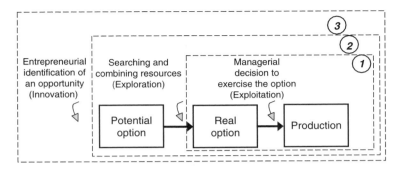

Figure 11.1 The option chain in the strategic literature

11.1.1 Standard option exploitation

The area labelled 1 on Figure 11.1 represents the initial step of the real option theory as it can be found in today's major textbooks on that topic (Amram and Kulatilaka, 1999; Trigeorgis, 1996). This short option chain is the following. The option is supposed to exist, to lie in the hands of the decision maker. The decision maker (always referred to as the 'manager' in this literature) evaluates the option contingently to the future possible states, and decides whether to exercise it or not. 'Exercise' means that the project enters a phase of building and operation. Once the firm is in an operational phase new options can arise such as expanding the size of the plant, diversifying production, or in the worst case, stopping production and shutting down the factory.

11.1.2 Exploration of the identification of opportunities, the emergence of the potential option

This development, as noted by academics in the strategic management field, makes the important assumption that the option exists and that the decision maker is informed of its existence. This is, obviously, not always the case. Bowman and Hurry (1993), struck by that implicit hypothesis, introduced the notion of a shadow option (or potential option as explained later on), the option that a firm could exploit, or at least consider in its portfolio of choices supposing it is aware of it. The option that a firm possesses depends on the resources and knowledge available inside the firm but ignored by decision makers.

Opportunities (and not yet options) come into existence when individuals have different beliefs in the possibilities offered by the available or potential resources to transform some inputs into specific outputs that can be sold and raise a profit (Kirzner, 1979). An opportunity is a favourable, momentary circumstance or situation that has been recognized after one has sought it or it has spontaneously appeared. Once the potential option is taken into account (area 2 in Figure 11.1) the rest of the option chain can be considered in the same way as described above. When the shadow option is recognized, it moves from a potential option label to become a real option. The option has then to be evaluated, and compared with other options; one must also take into account probable interactions between them in the portfolio of option decisions available to the firm.

The addition made by Bowman and Hurry (1993) does not completely answer the question of the origin of the real option; it merely shifts the debate. Instead of explaining the origin of the real option, the genesis of the potential option must be elucidated.

11.1.3 Introduction of the entrepreneurial resource

A historical approach of entrepreneurs in microeconomic theory (Barreto, 1989) shows that when authors need to introduce novelties or special variations into a theory of the firm they often refer to the figure of the entrepreneur. In this work we consider entrepreneurship as a resource of the firm and an opportunity exploiter (Cohendet et al., 2000). Kirzner introduced the concept of 'entrepreneurial alertness' as the special ability of the entrepreneur to see where products (or services) do not exist and can be profitably exploited. Alertness exists when one individual has an insight into the value of a given resource while others do not. From this perspective, entrepreneurial alertness refers to 'flashes of superior insight' that enable one to recognize an opportunity when it presents itself. As already mentioned, the starting point of the potential option is the identification of an opportunity.

Once the opportunity is identified one can note that the entrepreneur certainly does not have the specific knowledge and expertise in all domains necessary to fulfil his goal. He must still find and combine adequate resources for his endeavour. The acquisition of a new productive capability by building a new competence is not instantaneous. The building process is mainly path-dependent and involves the tacit knowledge acquired by learning by doing and experimentation. This implies that firms which create knowledge are also option-creating firms. By creating new knowledge these firms expand their cognitive frames, part of the real option heuristics. Here the value of the entrepreneurial resource appears as having the ability to combine different expert knowledge with the aim of exploiting opportunities. The decision to exercise the option and to turn to a production phase corresponds to a managerial decision-making process. The reason why the manager does not enter the option chain earlier comes from the nature of the output of the shadow option. The output of the knowledge-building process (what is done during the shadow option) is difficult to evaluate, knowledge is dispersed and the manager is not aware of all the pieces before the entrepreneur terminates his action. The managerial decision depends on the balance between entering the market with the actual resources and knowledge or waiting to absorb more capacities.

11.2 The model: creation of options and firms

This section details the model of creation of new options (discovery of technological alternatives) and the creation of new organizations. The formulations are directly inspired by the models of Miller and Arikan (2004), Nelson and Winter (1982), Winter (1984); Grebel et al. (2003) for the entrepreneurial representation. The market structure of the model is presented in Section 11.2.1; later we introduce a maximum level of development for each technology and then determine the alternatives (Section 11.2.2). In a third point we represent the creation of potential options by agents involved or external to an existing firm (Section 11.2.3).

Figure 11.2 schematizes the structure of the model. We distinguish three levels of analysis: the level of the agents, of the firm and of the industry. The entrepreneurial process creating the potential options begins at the agent level. An initial set of heterogeneous agents is generated. These agents are randomly paired. The gathered agents form a potential firm with a view to exploit an option discovered by one of them. This option is a new technology which enters the production process of a homogeneous good exchanged on the market. Their decision to build a firm is based on their joint evaluation of the economic situation which relies on the available economic indicators. If the agents decide not to found new firms, they return to the set of the potential agents for the following period. If the agents decide to found a firm, they are withdrawn from the set.

The firm is then established. The agents turn the potential option into real options that the firm can use to produce. At the industry level the new firm competes with existing firms. The competitiveness of the firm

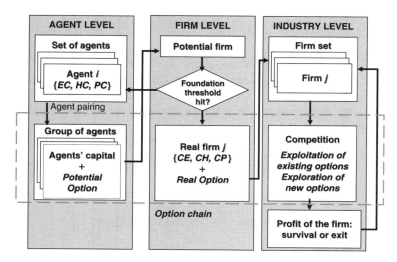

Figure 11.2 Interaction of the option chain with the agent, firm and industry

depends on the level of development of its technology and its stocks of physical and human capital. The process of selection that takes place is a competition over the quantity produced. At the same time as the competition for the development of existing technologies takes place, a competition for the exploration of new technologies occurs. Thus, within each existing firm, an entrepreneurial agent can at each period discover a new potential option. According to the reception reserved for this discovery by the firm, the option will be exercised within the firm and will increase the size of the technological portfolio of the firm. If the firm does not accept to develop the new technology a splitting of the firm into two entities takes place. One unit develops the new technology; the other focuses on the old techniques. The lengthy process of selection which takes place evaluates the entrepreneurial idea (the potential option), and the quality of the managerial decisions taken (transformation into real option) by the firms.

11.2.1 Market and industry

The market is represented by N firms producing the same homogeneous good. At the beginning of the simulation, each active firm F_i in the industry has two technologies, perfectly substitutable, with the same initial productivity and the same maximum capacity of development.

Each firm is determined by a quadruplet:

$$F_{i,t} = \{g, \{EC_{it}, HC_{it}, PC_{it}\}\} \quad (11.1)$$

In this expression g represents the gender of the firm, either an evolutionary firm, or an option-based reasoning firm. The gender is invariant for the firm over all the duration of the simulation. An evolutionary firm is a firm that only uses one technology as long as a satisfaction criterion is met; an option-based firm is a firm developing a portfolio of different options – see Miller and Arikan (2004) for a definition of those firms.

The elements EC, HC and PC respectively represent the entrepreneurial, human and physical capital (machines) of the firm. These elements are positive or nil. The combination of these three elements enables us to represent the creation of the potential and real options as well as the creation of new firms.

11.2.1.1 Demand and supply

The quantity of goods produced by a firm is Q_{it}. At each period, firms produce at the maximum of their capacity. The demand absorbs all the production emanating from the firms (with an adjustment of the prices); there is no limitation in terms of inputs of production as long as the firm is able to finance its expenditures. As in the model of Nelson and Winter, the total offer of the market is Q_t.

$$Q_{it} = T_{it}^* \cdot (q_h HC + q_p PC) \quad (11.2)$$

$$Q_t = \sum_{i=1}^{N} Q_{it} \quad (11.3)$$

where T^* represents the best technology used by the firm to produce. The HC element is an aggregate representing the whole human resources of the firm, both the agents in charge of production and the researchers. PC, the physical capital, corresponds to the machines used for production. The parameters q_h and q_p have several attributes. When the parameters are defined for the unit $[0; 1]$ they represent the share of agents and physical capital dedicated to the production. When the parameters are dependently defined ($q_h + q_p = 1$), they can be interpreted as the share of human capital necessary for production compared to the share of machines. Lastly, leaving aside these scales, they make it possible to calibrate the model. Price P_X for the homogeneous good is obtained by the following expression, identical to that employed by Nelson and Winter:

$$P_X = D(Q_t) = \frac{D}{Q^{1/\varepsilon}} \qquad (11.4)$$

with D the demand, and ε elasticity of the demand to the price.

The calculation of the sales turnover of the firm ($CA_{i,t}$) and profit is given by

$$CA_{i,t} = P_X \cdot T_{it}^*(q_h HC + q_p PC) \qquad (11.5)$$

$$\Pi_{it} = CA_{it} - c_e EC - c_H HC - c_p PC \qquad (11.6)$$

where the c's represent the various production costs per unit of capital.

The firms, who make a positive profit, invest it in R&D in order to develop technologies which they have in their portfolio and in their physical and human capital.

11.2.1.2 Investment in capital and R&D

We assume that the costs of acquisition of human, physical and entrepreneurial capital are constant. The profit released at the end of each period is allocated in the following way:

$$\Pi_{i,t} \equiv \Delta PC_{i,t} + \Delta HC_{i,t} + \Delta EC_{i,t} + S_{i,t} \qquad (11.7)$$

where $S_{i,t}$ is the capital invested in the improvement of existing technologies.

The variations of the different types of capital follow these expressions:

$$\Delta PC_{i,t} = \mathrm{Min}\left[0; \left(\frac{q_p}{q_h + q_p} - q_e\right)(\Pi_{i,t} - S_{i,t})\right] \qquad (11.8)$$

$$\Delta HC_{i,t} = \mathrm{Min}\left[0; \left(\frac{q_h}{q_h + q_p} - q_e\right)(\Pi_{i,t} - S_{i,t})\right] \qquad (11.9)$$

$$\Delta EC_{i,t} = \begin{cases} \mathrm{Min}[0; q_e(\Pi_{i,t} - S_{i,t})] & \text{if } Tn < TN \\ 0 & \text{otherwise} \end{cases} \qquad (11.10)$$

with

$$q_e = \frac{q_h}{2(q_p + q_h)}$$

Since our model aims at determining the effects of the creation and exploitation of technological options we give priority to the expenditure in R&D. The allocation of profit remaining after allocation to research depends on the choice made by the firm whether or not to develop several technologies in parallel. When Tn the number of alternative technologies the firm holds reaches TN (the maximum number of technologies it wishes to have in its portfolio), it stops investing in entrepreneurial capital; in the opposite case a remaining share of the profit is allotted to the entrepreneurial resource. When the profit is negative the firm sells its physical capital until it reaches a level of minimum capital which forces it to leave the industry.

After having clarified the mechanism of profit allocation, we now present the mechanism of alternative technology generation and development.

11.2.2 Determination of alternative technological options

We introduce the possibility of having TN technologies. Each technology allows the production of the same homogeneous good but with different potentials of development. A new technology is composed of two indications, the probability that this alternative is discovered and its maximum development potential. We model each one of these points according to several specifications, taking as a starting point those retained by Nelson and Winter.

11.2.2.1 Determination of the maximum productivity of a technology

The maximum level of development of an alternative technology, T_{n+1}, is computed according to two methods: (i) on the level reached by the firm and (ii) a technical progress which depends on the evolution of basic science with time. Whenever a new technology is given on the basis of the level reached by the firm, the maximum level is obtained by a random draw according to a normal law, so that

$$T_{n+1} \max = N(T_n^*; EC + HC) \qquad (11.11)$$

where T_{n+1} max represents the maximum level of development that technology $n+1$ can reach. The T_n^* variable has different significances depending on the case considered. In the case where the next technology is based on the maximum level of a firm, T_n^* represents the maximum level reached by the firm at the time when it seeks an alternative technology.

The parameters EC and HC are used to compute the standard deviation in the random drawing, so that $\sigma = (EC + HC)$. The summation of the two parameters can be interpreted as the ability of the entrepreneur to seek and combine (human) resources in order to give form to his creative idea.

In the second representation the maximum level a technology can reach is based on the assumption of a growth of knowledge underlying the technologies so that

$$T_{n+1} \max = N(\theta t^*; EC + HC) \quad (11.12)$$

Thus, the maximum level that technology $n+1$ can reach depends on the moment t^* where the random draw occurs and on a parameter θ representing the drift of the knowledge growth. The parameters $EC + CH$ have the same function as previously indicated. These specifications give the maximum level of development that a new technology can reach; this level is unknown to the entrepreneur and to the firm. It is only when the firm decides to develop a technology and that its development does not produce any further progress that it knows that the maximum limit has been reached (if there is no improvement during τ periods, the firm stops investing in that technology).

11.2.2.2 Creation of an option as a technological alternative

We retain several forms of options creation related to the nature of the firm and the nature of the option. The options can be potential or real, and can be created within an existing firm or by an entity resulting from the work of an entrepreneur (newly created firm or splitting of an existing one). Thus the creation of the options influences the dynamics of an industry since the entries and exits of firms on the market are directly dependent on them.

11.2.3 Creation of a potential and real option within an existing firm

The creation of a new potential option is represented as an entrepreneurial action based on the existing structures of the firm. It is the result of a random draw from a uniform law, such as

$$\text{Prob[creation of a potential option, } P_{PO}(t) = 1]$$
$$= \text{Min}[\omega(EC_t), 1] \quad (11.13)$$

where ω is a calibration parameter. The probability that the firm agrees to develop this potential option, to exercise it, in order to obtain a real option is given by

$$\text{Prob[exercise the potential option, } P_{RO}(t) = 1]$$
$$= \text{Min}\left[\frac{\Omega}{\omega}\left(\frac{HC_t + PC_t}{EC_t}\right), 1\right] \quad (11.14)$$

where Ω is a calibration parameter.

The creation of potential options, according to this description, corresponds to an innovation within the meaning of Schumpeter Mark I. In this approach the introduction of an innovation is independent from the human and physical capital of the firm (researchers and laboratory). A formulation describing the creation of a potential option as an innovation made by a laboratory (a large company) within the meaning of Schumpeter Mark II would be obtained by introducing into equation (11.13) the human capital HC which contains agents dedicated to R&D. An alternative solution would be the implementation in the model of an additional equation where the creation of potential options would depend on the human capital of the firm. However, we prefer the approach above since it respects the dichotomy between radical and incremental innovations (Abernathy and Utterback, 1978). Our description corresponds to a situation where laboratories ($HC + PC$) make innovations of the incremental type by developing real options. The radical innovations, the potential options, are then introduced by another team of researchers, those who have an entrepreneurial spirit.

The numerators and denominators of expression (11.14) are influenced by the investments in the various capital of the firm (EC, HC, PC). This ratio implies that a firm which obtained a potential option and has a low level of human and physical capital compared to entrepreneurial capital may refuse to exercise the potential option and does not develop the technology as an alternative. However, parameter EC depends on the number of alternative technologies TN which the firm wishes to have.

11.2.4 Creation of an option and creation of a firm

We describe a population of agents, among whom some will work out a potential option, who need to find resources (human and financial) with other agents to constitute a new firm, which will allow them to exploit the potential-real option.

11.2.4.1 Description of the agents

The agents are heterogeneous from the point of view of their initial endowment in physical capital and competences. The agents are composed of three elements, the entrepreneurial capital (*EC*), the human capital (*HC*) and the physical capital (*PC*). The entrepreneurial capital element determines the creative potential of the individual in the sense of the Schumpeterian entrepreneur. A higher value of this element increases the probability that the individual becomes an innovative entrepreneur and creates a potential option. The human capital corresponds to the stock of knowledge of the agent, both theoretical and practical. The physical capital represents his contribution if he decides to create a firm; it represents at the same time a contribution in liquidity, machines, buildings ... It results from this formulation that the absence of entrepreneurial capital excludes an agent from the creation of an option, that the absence of human capital does not enable him to develop this option, and that the lack of physical capital prevents him from creating his own firm. Also, to be able to exploit an option and to enter into competition the agents must acquire their capital.

Each variable of the triplet determines whether an individual agent belongs to the interval [0,1]. According to a cardinal scale, a higher value indicates a higher level of the element considered. An agent $a_{i,t}$ with $i \in \{1, ..., n\}$ is described at each moment t by a vector of the following form:

$$a_{i,t} = \{EC_{i,t}, HC_{i,t}, FC_{i,t}\} \qquad (11.15)$$

Each element of the triplet is obtained by a random drawing from a uniform distribution [0,1]. The initial population of agents, A_t, is obtained by generating n vectors a_i.

At each period a subset of k agents are randomly paired in order to evaluate their chances to create a firm together. For that the triplets of the agents are considered additive to obtain a potential $pf_{q,t}$. Formally:

$$pf_{q,t} = \begin{pmatrix} \sum_{i=1}^{k} ec_{i \in k, t} = EC \\ \sum_{i=1}^{k} hc_{i \in k, t} = HC \\ \sum_{i=1}^{k} pc_{i \in k, t} = PC \end{pmatrix} \qquad (11.16)$$

k is determined by a random draw from a uniform distribution limited by [2; *CE*max] where *CE*max corresponds to the maximum entrepreneurial capital observed in an active firm in the industry. The passage from a potential to a real firm depends on the characteristics of the potential firm and on the environment. For this the threshold ψ_t, based on the median value of stocks of capital of the firms present in the industry, must be exceeded. If the kth group of agents, which corresponds to the potential firm pf_q has a capital exceeding this threshold, the firm is created.

We compel the new firm to have only two technologies during a certain lapse of time, for two reasons: (i) to allow the firm to correctly enter into competition with the firms already present in the market, and (ii) since the firm was created with the aim of developing and exploiting a precise option, it would not be justified to seek alternatives immediately. If the threshold is reached the agent returns to the initial set which will be used during the next period to generate new potential firms.

11.2.4.2 Splitting of the firms

At the time of the development of a new option within a firm many divergences of opinions can appear. These divergences can lead a number of the individual agents of a firm to leave the organization to form a new one. In such a case a splitting takes place and various stocks of capital are divided to form two new entities. When the two firms split, each new entity is entitled with a set of technologies. One obtains the two best technologies in the portfolio, the other gets the newly discovered technology with the use of the potential option. The splitting takes place by drawing a random number X according to a uniform law [0; 1]. The first entity receives X per cent of the capital and the second the complement. The other parameters are identical for the two firms except for the technologies used.

11.2.4.3 Exit from the industry

The firms which are not sufficiently competitive leave the industry. This occurs when the physical capital falls under a minimum stock of capital threshold (PCmin) corresponding to the formulation retained by Winter (1984). Let us recall that when the firm makes negative profits it sells its physical capital. This negative profit can have two origins. It can come from the technology employed by the firm which is not sufficiently good to allow the firm to produce at a rather low average cost to compete with the other firms on the market. Or it can come from a exceedingly high stock of capital that generates costs higher than the sales turnover of the firm; this is in particular the case for firms with too much entrepreneurial capital. Thus, a firm which explores too many new technologies because of excessive entrepreneurial capital is rapidly forced to leave the market.

11.3 Results of simulations

This section offers an overall representation of the results of the model. The presentation is made according to the homogeneity or the heterogeneity of the firms compared to particular parameters. Importance is given to the distribution of indicators such as productivity, concentration and the profit of the firms.

11.3.1 Global evolution of the industry

We followed the protocol of Nelson and Winter (1982), with five starting situations corresponding respectively to markets with 2, 4, 8, 16 and 32 firms; a protocol of simulations of this type is used by other authors (Andersen, 2001; Yildizoglu, 2002). The number of firms retained also enables us to modify the percentage of firms of the evolutionary type and of the option type present at the beginning of the simulation. Five configurations were retained:

- Evo-opt0: 100% of evolutionary firms
- Evo-opt1: 75% of evolutionary firms, 25% of the option type
- Evo-opt2: 50% of evolutionary firms, 50% of the option type
- Evo-opt3: 25% of evolutionary firms, 75% of the option type
- Evo-opt4: 100% of firms of the option type

Case 1: All the firms are homogeneous

For each competing structure the firms are homogeneous at the beginning of the simulation in terms of stock of capital and technologies. The values chosen ensure that exploration–exploitation is the same in each configuration and that industry records the same technical progress in the first steps. Thus, the effect of the structure of the industry will only intervene during the simulation.

The first results relate the number of real options created during simulations and the firm entries and exits in the market. Figures 11.3(a)–(e) are average results observed on 100 simulations. The industry initial configuration (percentage of firms of a certain type) always affects the results. Figure 11.3(a) represents the number of options created by firms (already existing or created) during simulation. Two tendencies can be distinguished, according to the axis 'number of firms', and according to the axis 'composition of firms'. First of all, an increase in the number of firms involves an increase in the number of options; in addition, the increase in the proportion of firms with a reasoning based on options leads to a reduction in the options created. This is explained by the fact that

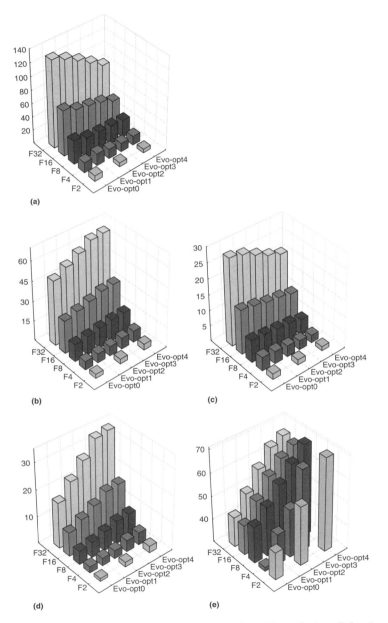

Figure 11.3 (a) Real options created; (b) total number of firms during all the simulation; (c) firms active at the end of the simulation; (d) firms created; (e) firms that exited the industry

option-type firms preserve their options for a longer time so that their portfolios of alternative technologies change slowly. The fact that option-type firms preserve their options at greater length leads the entrepreneurs who discover potential options within these firms to leave them and to create new entities. Figures 11.3(b) and (d) represent the total number of different firms which were active at one time during the simulation and the number of firms created by splitting or newly founded by entrepreneurs outside the industry. An increase in the option-type firms at the beginning leads to a clear increase in the creation of firms during the simulation.

Figure 11.3(c) represents the number of active firms at the end of the simulations and Figure 11.3(e) the number of firms who left the market. The exit rate is stable for the same ratio of evolutionary/option-type firms and increases with the proportion of firms of the option type. The smaller firms created to explore and exploit new options are more vulnerable than the larger firms installed which capitalize on a technology. The result from this is an increase in the exit rate from the industry.

These general results are obtained when all the firms are identical at the beginning of the simulation. Let us see how the results are modified when heterogeneity is introduced for some parameters.

Case 2: The firms are heterogeneous according to a characteristic

The modification of some characteristics also allows us to reconsider the interactions between the various elements of the model. The consequences of these modifications are presented below.

Stocks of human and physical capital. If the capital of a firm at the beginning of a simulation is higher than the capital of other firms, then this firm will produce more than the other firms. As a consequence Q, the total quantity offered, increases and P_X the price of the good drops compared to the basic case we presented. Since P_X is an element of the determination of the rate of profit this rate will fall. The stock of capital of all the firms will decrease accordingly. The same levels of P_X and Q as in our preceding simulations will be obtained but a firm will have a profit higher than the others, although the rate of profit is the same. The rapid fall in the price of the good leads to the exit of a greater number of firms from the market.

The costs of acquisition of capital. An increase of the costs for a firm induces a reduction of the capital it can acquire for the same amount of profit. This firm will not be able to acquire the same quantity of capital as the competitors and thus will not be able to produce as much. Conversely, a firm with lower costs will gradually dominate the market.

The starting technology. When the technology employed by a firm improves, its production and profits increase. Thus, a firm can acquire more productive capital and allocate more resources to its R&D to develop its technology again.

A notable advantage in one of these three characteristics leads the firm to dominate the long-term market.

11.3.2 Structure of industry at the end of the simulation

The following analyses are based on Figures 11.4(a)–(f). All these figures are obtained with an initial configuration of 32 firms but other configurations lead to close results. The representation of the box plots retained corresponds to the minimum and maximum for the whiskers and to the median for the point. For Figures 11.4(a) and (b) the value 100 corresponds to the starting situation of technologies. For Figure 11.5 the x-axis represents time and the ordinate corresponds to a stacking up of the results of 10 simulations, which makes it possible to appreciate the variations from one simulation to another.

11.4 Co-evolution of the options, industry and technologies

The preceding analysis allows us to conclude that the dynamics of the industry and the results of the firms are related to the characteristics of the options and technology. However, these analyses only took into account variations in one parameter. In the following we provide variations of several parameters. These parameters are fixed following the notion of technological regimes (Malerba and Orsenigo, 1993, 1995). First, we show the characteristics of each regime used (Section 11.4.1). Secondly, the results obtained for each regime are presented (Section 11.4.2).

11.4.1 Specification of the technological and market regimes

We retained two main categories of characteristics to distinguish the two technological and market regimes (TMR1 and TMR2), the technological characteristics and the characteristics related to the structure of the market. These characteristics as well as the scales of parameters and variables which we retained to specify them are summarized in Table 11.1.

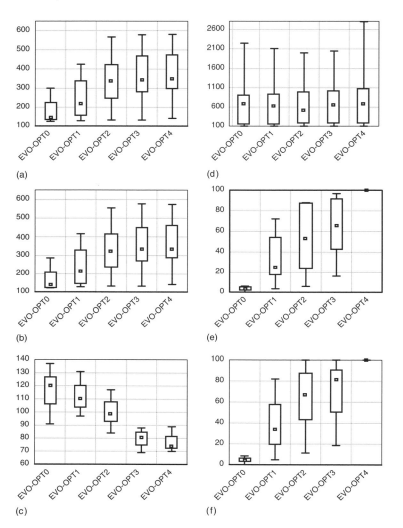

Figure 11.4 Box plots for 32 firms: (a) Average level of the technology; (b) maximum level of the technology; (c) number of alternative technologies; (d) average profit; (e) option-type firm profit part; (f) market share of option-type firm

11.4.1.1 Technological characteristics

We distinguish the two technological regimes according to five characteristics. These characteristics are related to the origin of the options (or the innovations), the transformation of the potential options into

Table 11.1 Specification of the technological and market regimes

Characteristics of the regime/ Parameters used	Technological and market regime 1 Entrepreneurial	Technological and market regime 2 Managerial
(a) Technological characteristics		
Origin of the options/ Entrepreneurial capital	Small entrepreneurs $Ce/CE > 1$	Big firms $Ce/CE < 1$
Transformation potential option to real option/ Ratio $(CH + CP/CE)$	Small CH and CP small	High CH and CP high
Effect of knowledge accumulation	Small	High
Learning by doing	$b1 = 1$	$b1 > 1$
Allocation of R&D credit	$b8 = 1$	$b8 > 1$
Appropriation of innovations	Small	High
Numbers of alternative technologies desired	$TN = 2$	$TN = 3$
Technological progress/ Maximum development potential of discovered technologies	Based on the firm	Based on fundamental science
(b) Characteristics of the market		
Firm entry rate on the market/ Entry threshold	High First quartile of capital stocks	Low Median of capital stocks
Size of entrant	Small entrepreneurs	Big firms
Number of agents needed to form an entrant firm	K small	K high
Concentration	Low concentration	Concentrated
Number, size and type of firm at $t = 0$	$N = 32$, 50% of firms are option type, all the same	$N = 32$, 50% of firms are option type, heterogeneous

real options (the use of technological opportunities), the importance of knowledge accumulation and the determination of the characteristics of the new technological alternatives.

Technological alternatives. We defined two modes of determination of the characteristics of new technologies: firm based or fundamental science based. In the regime we define below, the representation of the knowledge employed is different. Thus, for the regime TMR2, the evolution of knowledge is based on the development of fundamental science. The latter is supposed to be linear with time. The regime (TMR1) which corresponds to small entrepreneurial companies uses the maximum level of technology obtained by the firm to define the maximum potential of the technological options. This mode corresponds to a more practical than theoretical knowledge and requires a lower stock of knowledge for the creation of a new option.

Origin of the options. The potential options in our modelling can have two sources. They emanate either through entrepreneurs employed by an existing firm or not. As regime TMR1 is of an entrepreneurial type, agents not employed by firms have a stock of entrepreneurial capital higher than the stock of entrepreneurial capital of the active firms.

Transformation of the potential options into real options. When a new option is created by an entrepreneur, it seeks the physical and human capital which it lacks in order to turn the potential option into a real option. The change of option type within a firm depends on the relative stocks of the various types of capital, which constitute the opportunities, and on the facility offered to an entrepreneur to develop the option in-house. Mode TMR2 characterizes firms with large R&D laboratories which offer better conditions to develop the options in-house, by having a higher ratio $(CH + CP/CE)$.

11.4.1.2 Market characteristics

We distinguish the two regimes according to three characteristics related to the structure of the market at the beginning of the simulation: concentration, rate of entry and size of the firms entering the market.

Concentration. For the two regimes, the set of simulations is carried out with 32 firms, half of which are of the option type. Regime TMR2 is characterized by the existence of large companies; in addition, these firms have larger stocks of physical and human capital. These firms produce a greater part of the good exchanged on the market.

The rate of entry into the market and the size of the entering firms. Regime TMR1 corresponds to a more dynamic entrepreneurial structure than mode TMR2, with a large number of firms entering but of a smaller size. The entry is eased by reducing the threshold of necessary capital.

The next section analyses the results of the simulations made according to the defined regime modes.

11.4.2 Creation and evolution of the options in differentiated regimes

Figure 11.5 presents the results for each regime. To facilitate the reading and the interpretation of the results, we chose to present them expressed on the same base (100). Thus, the regime which leads to the smallest result according to the measure selected is taken as a basis of calculation to express the value obtained in the other regime.

11.4.2.1 Co-evolution of the structure of the industry and the options

In the two regimes the concentration is stronger at the end of the simulation than at the beginning. But the differential between the concentration at the beginning and at the end of the simulation is more important in regime 1. The number of active firms at the end of the simulation is higher in regime 1; this is explained by a larger amount of firm splitting (+89 per cent) and the number of entries of new firms. Lastly, the price is higher by almost 30 per cent in regime 1. The more stable structure of regime 2 creates less potential options, but develops them in a more effective way into real options. On average, the same technological option is used more than four times longer by the firms belonging to this mode. The best technologies are preserved and developed at greater length.

11.4.2.2 Co-evolution of technology and the options

There is a great disparity in the results concerning the maximum technology reached on average during the simulations. Regime 1 leads to twice more productive technologies, but obtaining such a technology is the prerogative of some firms only. So the average productivity of the technologies used by the firm is higher in regime 2 by almost 65 per cent. Regime 2 is based on continuous science development, while the entrepreneurial regime is marked by discontinuities in the form of technological jumps. These discontinuities, when they are repeated, thanks to favourable probabilities in the creation of the options, generate for regime 1 a fast technological progress, which explains the strong difference in the maximum productivity of technologies reached. The nature of knowledge, the turning-over of the options and the size of the technological portfolio condition the evolution of the productivity of technologies. The variation of one of these parameters modifies the average and maximum productivity observed.

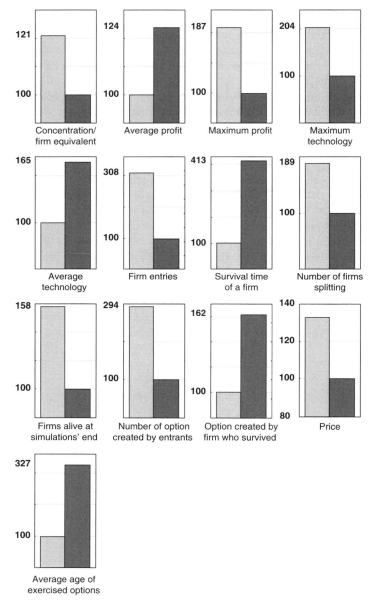

Figure 11.5 Technological and market regimes. Regime 1 (entrepreneurial) is always on the left, regime 2 (big firms/managerial) on the right. The smallest value of each is taken to express the second on a 100 basis

11.4.2.3 Co-evolution of the firms and the options

Observation of the indicators of Figure 11.5 related to the firm shows that regime 2 is accompanied by a higher profit (+24 per cent). The rate of survival of the firms is considerably higher in regime 2 and the number of splitting is lower. This is explained by higher capital allowing certain firms to produce more during the first period, leading to a fall in prices and eliminating the smallest firms quickly. The evolution of the firms, the growth of stocks of capital and the profits are related to the cycle of creation and exploitation of options. The firms of regime 2 develop them with more productive capital HC and PC than entrepreneurial capital.

11.5 Conclusion

The main results of the model are the following. First, the model shows that the firm that utilizes an option-based reasoning obtains higher profits than others and leads to a more advanced technology for the industry and a higher concentration of the industry. Secondly, the firms that dominate the industry are those that modify the proportion of entrepreneurial, human and physical capital between the creation step and the exercise of the option and those that master the different steps of the option chain. In this view, this approach develops and completes the debate on the duality between entrepreneur and manager in evolutionary theory (Cohendet et al., 2000).

The results presented in this section highlight the determining role of the characteristics of the technological environment and the structure of specific industries belonging to each regime on the creation and the development of options. This effect is not one-sided, since options determine in their turn future technologies and the evolution of the firms. Initially, we focused on the links between the options and the structure of the industry which enable us to detect the importance of the entries of new firms and their effects on production and the level of price. In a second step, we were interested in the links between the nature of knowledge, technology and the options. We observed the existence of a double effect which these parameters have on the options, both on the creation of new options and the development of the existing ones. Lastly, in the third phase we noted the existence of a co-evolution between the profits released by a firm, its composition compared to the various types of capital and its options.

References

Abernathy, W. and J. Utterback, 'Patterns of Industrial Innovation', *Technology Review*, June–July (1978), 41–7.
Adner, R. and D. Levinthal, 'What is not Real Option: Identifying Boundaries for the Application of Real Options to Business Strategy', *Academy of Management Review*, XXIX (2004), 74–85.
Amram, M. and N. Kulatilaka, *Real Options: Managing Strategic Investment in an Uncertain World* (Cambridge, Mass.: Harvard Business School Press, 1999).
Andersen, E. S., 'Toward a Multiactivity Generalisation of the Nelson–Winter Model', Druid Nelson and Winter Conference, 12–15 June, Aalborg, Denmark (2001).
Barreto, H., *The Entrepreneur in Microeconomic Theory: Disappearance and Explanation* (London: Routledge, 1989).
Bowman, E. H. and D. Hurry, 'Strategy through the Option Lens: an Integrated View of Resource Investment and the Incremental-Choice Process', *Academy of Management Review*, XVIII (1993), 760–82.
Cohendet, P., P. Llerena and L. Marengo, 'Is there a Pilot in the Evolutionary Theory of the Firm?', in N. Foss and V. Mahnke (eds), *Competence, Governance and Entrepreneurship* (New York: Oxford University Press, 2000), pp. 95–115.
Cyert, R. and J. G. March, *A Behavioral Theory of the Firm* (Englewood Cliffs: Prentice Hall, 1963).
Grebel, T., A. Pyka and H. Hanusch, 'An Evolutionary Approach to the Theory of Entrepreneurship', *Industry and Innovation*, X (2003), 493–514.
Kirzner, I. M., *Perception, Opportunity and Profit* (Chicago: University Press of Chicago, 1979).
Levinthal, D. and J. G. March, 'A Model of Adaptive Organizational Search', *Journal of Economic Behaviour and Organization*, II (1981), 307–33.
Malerba, F. and L. Orsenigo, 'Technological Regimes and Firm Behaviour', *Industrial and Corporate Change*, II (1993), 45–72.
Malerba, F. and L. Orsenigo, 'Schumpeterian Patterns of Innovation', *Cambridge Journal of Economics*, XIX (1995), 47–65.
Miller, K. D. and A. T. Arikan, 'Technology Search Investments: Evolutionary, Option Reasoning, and Option Pricing Approaches', *Strategic Management Journal*, XXV (2004), 473–85.
Nelson, R. R. and S. G. Winter, *An Evolutionary Theory of Economic Change* (Cambridge, Mass.: Harvard University Press, 1982).
Penrose, E. T., *The Theory of the Growth of the Firm* (Oxford: Oxford University Press, 1959).
Teece, D. J., G. Pisano and A. Shuen, 'Dynamic Capabilities and Strategic Management', *Strategic Management Journal*, XVIII (1997), 509–33.
Trigeorgis, L., *Real Options: Managerial Flexibility and Strategy in Resource Allocation* (Cambridge, Mass.: MIT Press, 1996).
Winter, S. G., 'Schumpeterian Competition in Alternative Technological Regime', *Journal of Economic Behaviour and Organization*, V (1984), 287–320.
Yildizoglu, M., 'Competing R&D Strategies in an Evolutionary Industry Model', *Computational Economics*, XIX (2002), 51–65.

12
Financial Means' Competencies and Innovation: Comparative Advantages between SMEs and Big Enterprises

Francis Munier

12.1 Introduction

The interpretation of Schumpeter's works (1935, 1939, 1974) leads to the formulation of two major assumptions. The first suggests that a positive relation between an innovation and the power of the monopoly prevails. The second supposes that big companies innovate more proportionally than small and medium-sized enterprises (SMEs) (the intensity of the research increases more than proportionally with size). This conjecture has been analysed in several empirical studies (Kamien and Schwartz, 1975, 1982; Baldwin and Scott, 1987; Scherer, 1992). These works brought answers, but they also show, as recalled by Le Bas (1991), that this question does not present either evident theoretical or empirical proof. According to Scherer (1992), an accepted result would be that big companies are more predisposed to innovate since they have greater means (financial, human, etc.), while the advantage of the small firm in the process of innovation is especially focused on the organizational level. This result led some authors to propose other research. Cohen (1995) notably suggests targeting their work to the concept of competence to study the explanatory factors of the innovating behaviour rather than to try to analyse a direct relation between size and innovation, as an important source of bias (Acs and Audretsch, 1990).

The aim of this chapter consists in introducing this dimension of 'competence' into the study of the conjecture and to verify whether big companies possess more competencies in terms of means to innovate. The interest is to give a wider meaning to the notion of R&D in order 'to sweep' a bigger range of innovative behaviours, notably those of small and medium-sized firms. Our empirical work bases itself on the exploitation of a database stemming from a French inquiry by the Service of Statistics and Industrial Studies of the French Ministry of the Economy (SESSI) in 1997. We estimate possession of competencies to innovate in French industry by distinguishing three classes of size: from 20 to 99 employees (a small firm), from 100 to 499 employees (a medium-sized company) and over 499 employees (a big company). We take into account sector-based effects by determining four groups according to the intensity of the sectors. The empirical analysis bases itself on a logit model determining the probability of having a competence according to the class of size.

The first section of this chapter is dedicated to a brief review of the empirical literature about the explanatory factors of the relation between the size of the firm and the means. The second section describes our empirical analysis. We successively present the database, the methodology and the econometric model.

12.2 Review of the empirical literature

Here we deal with the questions connected to the costs of research, the economies of scale, and the financial and investment means according to the size of the company. We then conclude with the methodological limits of the empirical approaches as well as with a reminder of some salient facts concerning the Schumpeterian conjecture.

12.2.1 High level of fixed cost and indivisibility of research

The existence of high R&D fixed costs is obviously a handicap for SMEs, in particular due to a gap between the rate of return and the level of sales expected (Cohen and Klepper, 1996). This is connected to a central point in Schumpeter's model: big companies have a greater market power.

This commercial power is the mainspring of innovative activity of companies since they can finance research with greater expected profits and because the appropriation of the innovation will be bigger. Size then appears as a restrictive factor, in particular if we presume that companies run their innovations by means of their own production. Few empirical studies were carried out to try to determine if the costs of R&D are systematically brought up. On the other hand, it is relevant to consider that in the case of a high research cost, big companies are favoured because of bigger financial means and a greater inclination to diversify their innovative activities (diversification of the risks on a portfolio of projects).

In addition, it is undeniably more relevant to consider the existence of an indivisibility of the R&D in certain industries than to refer systematically to the high cost of the research (Cohen and Klepper, 1996).

12.2.2 Economies of scale in the production of innovations

A number of scholars mention the existence of economies of scale in the production of innovation in order to justify the advantage of the big company (Galbraith, 1956). The question is to know if the relationship between the number of innovations and the expenses of R&D (or workforce dedicated to the R&D) increases with the size of the company. Empirical studies do not allow us to prove the existence of economies of scale in the production of the innovation. The main result would rather be the existence of a decreasing productivity with the size of the company. We can quote the study led by Acs and Audretsch (1991), which shows that returns to scale decrease or remain constant with regard to the innovation in 14 sectors (the most representative) of the American economy. The inquiry of Pavitt et al. (1987) ends with a more contrasting result, by showing that globally the relationship of the number of innovations in the expenses of R&D decreases with the size of the company, although for certain sectors the relation presents a 'U' shape. In the same way, Bound et al. (1984) conclude that the production of patents on the research expenses is higher for small and big companies than for average companies.[1]

Since R&D productivity is measured on the basis of formal research statistics, it appears that the contribution of small firms is underestimated. In the same way, Mairesse and Cunéo (1985) consider that companies that do some research are not necessarily the most successful because other efficiency factors come into play. This result is close to the remark of Cohen and Klepper (1996), who suggest that the existence of a strong profitability of the research does not necessarily confer an advantage (or a disadvantage) from the point of view of innovation.[2] The productivity of R&D according to size can also be examined from the point of view of the economic value of the innovation. Indeed, the number of innovations produced does not matter if these have, in fact, no monetary value (for innovations of products). The observations made in this respect do not bring a definitive answer since it is difficult to quantify the value of the innovation, thanks notably to the numerous incremental improvements it undergoes during its evolution. Moreover, the rate of return is different according to the nature of the research (basic, applied or developmental).

After all, and despite the limits inherent in samples and in the measurement of economies of scale, all empirical work seems to indicate that the productivity of R&D decreases with size. Nevertheless, Cohen and Klepper (1996) suggest that this apparent contradiction enters a growing linear function, even more than proportional, between size and R&D. Decreasing R&D productivity is explained by the phenomenon of the spreading of research costs among the various units.

12.2.3 Financial capacities and investments

The superiority of the big company also rests on the existence of greater financial means. In order to empirically verify this assertion, various hypotheses must be tested. Do big companies have proportionally higher internal financial means? Is access to external financing on the stock and banking exchanges easier for big structures? In addition, is it established that financial constraints have an effect on innovative activity? In other words, is it necessary to reject the theorem of the separation between financial means and decisions to invest? The hypothesis rests on the fact that financial constraints influence the size–innovation relation. In big companies, having a market power confers possibilities of internal financing and easier access to external financing. The distribution of risks is also possible, which brings us back to the argument of indivisibility of research opportunities and thus to the effect of threshold of size.

Empirically, SMEs are characterized by financial means less than those of big companies. Stiglitz and Weiss (1981) show that rationing credit is dependent on the size of the firm. Fazzari et al. (1988) reach the following result: liquid assets are influenced by size in a positive way. Evans and Jovanovic (1989) underline that financing constraints are bigger for small companies. It is also the case for French companies; Levratto (1994) considers that the financial constraints of SMEs are articulated around four points: a lack of stockholders' equity, a short-term excessive debt, an excessive weight of inter-firm credits and large disparities in fixing a price scale for loans.

The first studies did not allow establishment of a significant positive relation between the volume of liquid assets and innovation (Kamien and Schwartz, 1982). On the other hand, recent studies seem to confirm the incidence of financial constraints on innovative behaviour, invalidating the theorem of separation. The inquiry led by Fazzari et al. (1988) clearly shows that financing constraints have an effect on the investments made by the company. In addition, Acs and Isberg (1991) show that innovation (represented by specific assets) is connected to the structure of the capital of the company. By applying the model of Evans and Jovanovic (1989), Levratto (1994) shows the major incidences of a financial incapacity on the innovative activity. Innovation is thus influenced in a positive way by the availability of stockholders' equity. The capacity to self-finance therefore plays a dominant role at the level of commitment to research. The profitability of innovation is positively connected to the capacity of self-financing in the sense that R&D has a higher cost because credits are more expensive. Moreover, a limited capacity of self-financing forces the company to decrease dividends, which can be a source of conflict between shareholders and managers. This possible conflict later disrupts the strategies of the firm and thus then decreases the expected performance.

Most of the empirical investigations confirm that the incidence of financial constraints on innovative activity lessens with size. Other studies nevertheless showed that the relation between liquidity and innovation has an inverted 'U' shape according to size (Antonelli, 1989). This result can be interpreted by the fact that small firms invest more in a regime of strong competition to develop a competitive advantage (if they have the possibility). Moreover, criticism can be formulated concerning the interpretation of the influence of liquidity. Cohen (1995) clarifies that this influence can be understood according to two different points of views: *ex-ante* or *ex-post* with regard to the innovative investment. This precision is important according to the criteria of choice of the investment but does not contradict the report of the financial weakness of small firms. It is also relevant to take into account the multiple public funds granted to SMEs in order to help them to overcome the weight of a research project. Globally, all public funds contribute to minimize the financial constraint in the sense that they strictly speaking relieve the financing, as well as the cost of training and expert evaluations, which costs are sometimes overstated for SMEs.

In conclusion, the limits inherent in empirical analyses lead to a multiplicity of results, sometimes diverging. In general, the most striking results show that the intensity of R&D increases in a regular way with the company's size. However, some studies clarify that size has a negative effect in terms of research inputs, and subsequently for very big sizes, that the relation becomes positive ('U' curve). Besides, the relationship between R&D and sales varies more according to sector than according to the strength observed in the same sector. More precise empirical analyses considering the sector-based effects confirm the existence of a proportionality between size and R&D.

We can retain the idea, according to which big companies put more means into the preparation of the innovative activities (notably due to the existence of greater financial means and thanks to a bigger diversification). Big companies are generally more able to display a formal R&D. They have relative opportunities in the process of appropriation, for example in patents (in financial terms and in terms of knowledge). Nevertheless, the probability of implementing an innovative activity remains high for small companies in many sectors. This is explained by the fact that small firms operate by informal research. The measure in terms of R&D expenses according to official international definitions considerably hides this aspect. The second major result of the various empirical works is that the production of innovation increases less than proportionally with the size of the firm. In other words, the contribution of small firms is greater than in the case of big companies. The third important result is that R&D productivity increases less than proportionally with the company's size. In sum, we may conclude from this that the explanatory mechanisms of the size–innovation relation depend more on the organizational features of the firm.

Acs and Audretsch (1990) suggest that it is no longer so much the study of a functional relation between size and innovation that is important but the exploration of the determinants of the distinctive innovative behaviour of companies according to their size. Innovation is not an activity that is reduced to the existence of a laboratory; it is on the contrary a transverse activity in the company that often requires cooperation with other services (marketing, financial, technical, etc.). There are nevertheless empirical difficulties in observing all the explanatory factors. A way of bypassing these limits would be to give a generic dimension to these mechanisms. We therefore suggest analysing the Schumpeterian conjecture from the angle of the concept of competence to innovate.

12.3 Empirical analysis

12.3.1 Presentation of the database

The database results from an investigation carried out by the SESSI during 1997. It was carried out on a sample of 5000 French industrial enterprises with more than 20 employees. The rate of answer was 83 per cent in number of units and over 95 per cent in terms of turnover. The sample, which we exploited initially, includes 3715 firms (see Munier, 1999a, b; François et al., 1999 and SESSI, 1998a, b for other analyses).

Using the terminology suggested by the SESSI, firms answered a questionnaire related to the possession of 73 competencies ('elementary competencies') gathered around 9 'complex competencies'. The 73 competencies constitute the total competence of the company. Competencies measured in the investigation are competencies at the level of the firm. The point is to know whether or not the company has competencies related to the process of innovation.

Despite the richness of the database, criticisms can nevertheless be formulated on at least four points. First, the investigation does not make it possible to know if the firms interviewed consider that a given competence is truly necessary in its own case or generally, to develop an innovation. Second, some competencies are not specific to innovation. It is then difficult to determine the objectives for which the firms developed these competencies. Generally speaking, the question of the sources of competencies is not treated; only the possession of a repertory of competencies at a given point in time is required. Third, insofar as the answers are binary (the questions relate only to the self-declared possession of competence, without reference to the position of the company in terms of competition), a direct comparison between two firms of different size, both having a given competence, proves to be difficult. Fourth, the qualitative nature of the data may be at the origin of certain biases. The results obtained rest on a self-declaration by the people interviewed who can be tempted to answer in a favourable way to give a positive image of the firm. Moreover, the official framework of questionnaires relating to research seems to support, a priori, large firms (Kleinknecht, 1987). In fact, considering known biases the answers reflect an opinion or a perception more than a measurable reality. We can nevertheless moderate these criticisms with at least three specific characteristics of our investigation. On the one hand, the interest of our database is precisely that it offers SMEs an opportunity to answer just as large companies do insofar as innovation is defined there in its broadest meaning. In addition, the questionnaires were filled out by people occupying similar functions in the firms (head offices, technical staff). Moreover, the importance of our sample tends to consider that biases of rather optimistic firms can be counterbalanced by answers of other rather pessimistic companies.

Despite these criticisms, the database remains highly valuable. It is, indeed, remarkable to be able to have such statistics, which provide detailed information on the bases of the innovation of the firms (see Figure 12.1).

12.3.2 Methodology

We propose examining the possession of means and financial competencies to innovate according to three classes of size: from 20 to 99

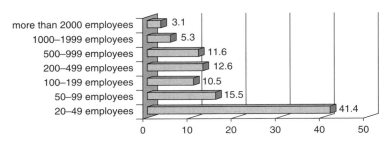

Figure 12.1 Distribution (in %) of the companies according to the class of size
Source: SESSI (1997)

Table 12.1 Means and financial competencies

Competencies	Variable
R&D	comp504
Do you subcontract or outsource R&D?	comp505
Do you use the inventions of others (patents, licences)?	comp508
Do you recruit highly qualified, scientific employees in order to innovate?	comp509
Have you purchased parts of companies or entire companies in order to innovate?	comp510
Do you think through all the costs of an innovative concept ahead of time?	comp801
Do you evaluate the cost of past innovations?	comp802
Do you know the private and public means for funding innovation?	comp803
Do you provide potential sources of innovation funding with information?	comp804
Do you have a specific promotional strategy for the new product?	comp901
Do you determine the target, media and type of publicity for the new product?	comp902
Do you project your company as being 'innovative and avant-garde' (offices, communications, publications)?	comp903

Source: SESSI (1997).

employees, from 100 to 499 employees and 500 employees and more. This regrouping makes it possible respectively to treat the empirical differences between the small, the medium and the large companies and thus widens the traditional opposition between SMEs and large firms.

Table 12.1 gives means and financial competencies, which we analyse.

To take into account the sectoral effects in terms of technological intensity, we refer to the nomenclature suggested by the OECD. Classification

Table 12.2 Distribution (in %) of the companies according to their size in the sectors by technological intensity (except energy)

Industrial sectors according to technological intensity	Small companies	Medium-sized companies	Large companies
High technological intensity (HT) (199 companies)	28	22	50
Average high technological intensity (MHT) (482 companies)	37	29	34
Average low technological intensity (MFT) (821 companies)	43	27	30
Low technological intensity (FT) (687 companies)	54	28	18

Source: SESSI (1997).

is carried out according to the intensity (direct or indirect) of the R&D after weighting the 10 main Member States. Four groups are thus constituted: sectors of high technological intensity (HT), of average high technology (MHT), of average weak technology (MFT) and of weak technology (FT). In these regroupings, we only consider the firms innovating in terms of products and/or processes. Consequently, the sample is reduced to 2189 companies. Statistically, the sectoral distributions are shown in Table 12.2.

12.3.3 The model

The model that we examine consists in estimating the probability of holding a competence according to the class of size (small, medium or large company).

The variable $comp_i$ represents a competence such as

$$P_i = P[\text{comp}_i = 1] = F[\beta_{T_{PE}} T_{PE} + \beta_{T_{ME}} T_{ME} + \beta_{T_{GE}} T_{GE}] \quad (12.1)$$

with $i = 1, \ldots, 12$ representing elementary competencies and where F indicates the function of distribution of a known law of probability. $\beta_{T_{PE}}$, $\beta_{T_{ME}}$ and $\beta_{T_{GE}}$ represent respectively the estimated coefficients of the variables (small, medium and large firms). All the variables are dichotomy variables. Value 1 for the competence variable indicates that the company has this competence. Concerning the size of the firm, their distribution into classes means that if a variable takes value 1, the others are worth 0.

In more condensed form, the model is written: $P_i = P[\text{comp}_i = 1] = F(\beta' T_j)$ with j representing three classes of size and $\beta' = (\beta_{T_{PE}}, \beta_{T_{ME}}, \beta_{T_{GE}})$.
We consider that the distribution of probability F follows a logistic law. Equation (12.1) thus defines the logit model.

The estimator of the β coefficients is obtained by the method of the maximum likelihood function. The likelihood function of the model is written:

$$l = \prod_{i=1}^{n} \left\{ P_i^{\text{comp}_i} (1 - P_i)^{1-\text{comp}_i} \right\} \tag{12.2}$$

By taking the log of l, one obtains the function log-likelihood L that is maximized compared to β:

$$L = \sum_{i=1}^{n} \left\{ \text{comp}_i \ln F(\beta' T_j) + (1 - \text{comp}_i) \ln F(-\beta' T_j) \right\} \tag{12.3}$$

The function is strictly concave, which makes it possible to ensure a likelihood maximum single for the logit model.

For each regression, we calculate the marginal effect relating to a class of size. This expressed as a percentage measures the increase (or the reduction) in probability of having competence if the company is respectively small, medium or large. The marginal effect is obtained by calculating the first derivative of the expected mean $E(\text{comp}_i) = F(\beta' T_j)$ compared to T_j:

$$\partial E \left(\frac{\text{comp}_i}{\partial T_j} \right) = f(\beta' T_j) \beta \tag{12.4}$$

where f represents the density of the law of probability.

Within the framework of the logit model where the function of distribution is of the following form:

$$F(\beta' T_j) = \frac{1}{(1 + \exp(-\beta' T_j))} \tag{12.5}$$

the marginal effect of the classes of size is calculated in the following way:

$$\frac{\partial E(\text{comp}_i)}{\partial T_j} = \frac{\exp(\beta' T_j)}{(1 + \exp(\beta' T_j))^2} \beta \tag{12.6}$$

Insofar as the marginal effect depends on the values on T_j, we use in calculation the average values of the variables.

12.3.4 Results and comments

The R&D competence makes reference explicitly to the capacities of the company to perform R&D, use the inventions of a third party, subcontract or acquire R&D and recruit highly competent staff (e.g. a researcher) (respectively, it is about competencies comp504, comp508, comp505, comp509) (see Table 12.3). We observe that the competence of companies to perform R&D is relatively important in all the branches of industry. Globally, big companies are the most concerned and average companies to a lesser extent. Small firms in the MFT and FT sectors are rarely competent. This result is interesting in that it confirms previous studies showing the superiority of big companies in R&D.

Big companies in the HT and MHT sectors are more competent in R&D management with third parties. In other branches of industry, these competencies are seldom represented whatever the size of the company. We can underline that it is not so much size that matters but more membership in a group of sectors where technological intensity is important. Performing R&D makes it possible to draw competencies from the appropriation of external resources. This competence is less developed by small-sized companies, even in technological sectors. Absorption capacity (Cohen and Levinthal, 1990) is more present in big companies. We find similar observations concerning the hiring of researchers, a variable strongly correlated to R&D.

The 'financing innovation' competencies concern the capacity of the company to deal with the costs of the innovation, to know the sources of financing, to find potential sources and to buy companies to innovate (respectively the competencies comp801, comp802, comp803, comp804 and comp510) (see Table 12.4). Thus we do not directly analyse financial means but more related factors. As regards financing, companies have specific competencies to anticipate the costs of the innovation and identify sources. For these two competencies, big companies are the most competent in all the branches of industry. The probability of having these skills decreases with technological intensity; in addition, marginal effects also decrease, in particular for small firms. On the other hand, the skills linked to the evaluation of the costs of the innovation and communications about potential sources are not representative. The difficulty of estimating the costs of innovation is obvious for many companies. Globally, traditional measures of the costs of the innovation concern only a part of innovation process. A study led by the BETA (1995) shows

Table 12.3 Estimation results for 'R&D' competencies

Competencies	Estimator of $\beta_{T_{PE}}$	Estimator of $\beta_{T_{ME}}$	Estimator of $\beta_{T_{GE}}$	Marginal effect $\beta_{T_{PE}}$	Marginal effect $\beta_{T_{ME}}$	Marginal effect $\beta_{T_{GE}}$	Probabilities
High-intensity sectors							
comp504	0.98	1.67	2.75	0.10	0.17	0.28	0.88
	(3.24)*	(4.04)*	(6.53)*				
comp505	−0.41	0.46	0.99	−0.10	0.11	0.23	0.62
	(−1.47)	(1.49)	(4.42)*				
comp508	−0.80	−0.09	1.27	−0.19	−0.02	0.30	0.60
	(−2.76)*	(−0.30)	(5.24)*				
comp509	−0.33	0.98	1.66	−0.07	0.20	0.33	0.72
	(−1.21)	(2.90)*	(6.08)*				
Medium high-intensity sectors							
comp504	1.16	2.06	2.18	0.14	0.25	0.27	0.86
	(6.49)*	(7.76)*	(8.53)*				
comp505	−0.50	−0.20	0.83	−0.12	−0.05	0.21	0.51
	(−3.17)*	(−1.17)	(4.95)*				
comp508	−0.70	−0.52	0.69	−0.17	−0.13	0.17	0.46
	(−4.33)*	(−2.99)*	(4.24)*				
comp509	−0.65	0.17	0.77	−0.16	0.04	0.19	0.52
	(−4.05)*	(1.01)	(4.67)*				
Medium weak-intensity sectors							
comp504	0.54	1.81	2.75	0.08	0.26	0.40	0.83
	(4.86)*	(9.36)*	(10.33)*				
comp505	−0.87	−0.15	0.29	−0.21	−0.04	0.07	0.42
	(−7.46)*	(−1.14)	(2.27)*				
comp508	−1.20	−0.59	0.16	−0.27	−0.13	0.04	0.35
	(−9.47)*	(−4.18)*	(1.26)				
comp509	−1.44	−0.55	0.13	−0.32	−0.12	0.03	0.33
	(−10.60)*'	(−3.92)*	(1.01)				
Weak-intensity sectors							
comp504	−0.41	1.03	1.39	−0.10	0.25	0.34	0.58
	(−3.83)*	(6.31)*	(6.20)*				
comp505	−1.43	−0.64	0.08	−0.29	−0.13	0.02	0.28
	(−10.83)*	(−4.23)*	(0.44)				
comp508	−1.85	−0.76	−0.08	−0.33	−0.13	−0.01	0.23
	(−12.16)*	(−4.91)*	(−0.44)				
comp509	−2.05	−0.78	0.08	−0.34	−0.13	0.01	0.21
	(−12.50)*	(−5.04)*	(0.44)				

Figures between brackets are the *t*- Student. The values with * are the critical value for 5% non-significant.
Source: SESSI (1997).

the importance of the transversal costs of innovations. The idea consists in considering that the process of innovation takes place in a transverse dimension of the activities articulating the elaboration and the management of the new knowledge. An activity-based costing then seems

Table 12.4 Estimation results for financial competencies

Competencies	Estimator of $\beta_{T_{PE}}$	Estimator of $\beta_{T_{ME}}$	Estimator of $\beta_{T_{GE}}$	Marginal effect $\beta_{T_{PE}}$	Marginal effect $\beta_{T_{ME}}$	Marginal effect $\beta_{T_{GE}}$	Probabilities
High-intensity sectors							
comp801	0.48	0.66	1.66	0.09	0.12	0.31	0.75
	(1.74)	(2.07)*	(6.08)*				
comp802	−0.04	−0.18	−0.41	−0.01	−0.04	−0.10	0.44
	(−0.13)	(−0.60)	(−1.99)*				
comp803	0.56	0.56	1.32	0.11	0.11	0.27	0.72
	(2.00)*	(1.79)	(5.40)*				
comp804	−0.56	−0.56	−0.16	−0.14	−0.14	−0.04	0.41
	(−2.00)*	(−1.79)	(−0.80)				
comp510	−2.10	−0.76	−0.53	−0.41	−0.15	−0.10	0.27
	(−4.86)*	(−2.35)*	(−2.57)*				
Medium high-intensity sectors							
comp801	0.50	0.55	1.01	0.11	0.12	0.22	0.67
	(3.17)*	(3.15)*	(5.77)*				
comp802	−0.35	−0.52	−0.49	−0.08	−0.12	−0.12	0.39
	(−2.28)*	(−2.90)*	(−3.03)*				
comp803	0.47	0.77	0.89	0.10	0.17	0.20	0.67
	(3.02)*	(4.25)*	(5.22)*				
comp804	−0.62	−1.01	−0.49	−0.14	−0.22	−0.11	0.33
	(−3.90)*	(−5.31)*	(−3.06)*				
comp510	−2.21	−1.70	−0.64	−0.33	−0.25	−0.09	0.18
	(−8.65)*	(−7.31)*	(−3.94)*				
Medium weak-intensity sectors							
comp801	0.31	0.59	0.98	0.07	0.13	0.23	0.64
	(2.87)*	(4.18)*	(6.93)*				
comp802	−0.22	−0.38	−0.21	−0.05	−0.09	−0.05	0.44
	(−2.35)*	(−2.74)*	(−1.73)				
comp803	0.25	0.83	1.22	0.06	0.18	0.27	0.67
	(2.35)*	(5.69)*	(8.09)*				
comp804	−0.70	−0.73	−0.72	−0.15	−0.16	−0.16	0.33
	(−5.94)*	(−13.05)*	(−6.12)*				
comp510	−2.11	−1.81	−1.22	−0.26	−0.23	−0.15	0.15
	(−12.25)*	(−9.36)*	(−8.09)*				
Weak medium-intensity sectors							
comp801	0.15	0.29	0.83	0.04	0.07	0.20	0.58
	(1.46)	(2.00)*	(4.26)*				
comp802	−0.50	−0.21	−0.27	−0.12	−0.05	−0.07	0.41
	(−4.64)*	(−1.44)	(−1.52)				
comp803	−0.14	0.51	0.75	−0.04	0.13	0.19	0.55
	(−1.35)	(3.41)*	(3.93)*				
comp804	−1.03	−0.85	−0.91	−0.21	−0.17	−0.18	0.28
	(−8.68)*	(−5.40)*	(−4.57)*				
comp510	−2.13	−1.91	−1.24	−0.24	−0.22	−0.14	0.13
	(−12.59)*	(−8.92)*	(−5.79)*				

Figures between brackets are the *t*- Student. The values with * are the critical value for 5 % non-significant.
Source: SESSI (1997).

more appropriate than standard cost accounting if we want to take into account all the costs connected to a project. There are nevertheless limits connected to the implementation of such an accounting, which seems to interest the big companies more (or the average companies). The interest of this study is especially to show how the inputs of innovation are difficult to measure in an exhaustive way and to underline the transverse dimension of innovation. We also observe that financial means connected to the purchase of the other companies with the aim of innovating are seldom representative for French industrial firms. This observation allows us to qualify the nature of the capacity of appropriation of big companies which expresses itself more in terms of R&D relations.

The 'sale of innovation' competencies express the capacity of the company to promote and valorize its innovation on the market (comp901, comp902, comp903). The competencies relative to sales techniques (comp901 and comp902) have a rather variable frequency. Companies do not show a lot of skills in the implementation of a specific promotional strategy concerning new products (comp901), but they know how to target their prospective customers (comp902) (see Table 12.5). Big companies are the most competent in sectors HT, MHT and MFT (in sectors FT, the observation is not significant).

The competence allowing an innovative image of the company to be given concerns more the big companies, whatever the branch of industry. Let us note that the probability that companies implement this form of strategy is stronger in sectors with a strong degree of technological intensity. This observation underlines the managerial and commercial deficit of small firms.

It seems that globally French companies have not enough skills in domains related to innovation. In particular, SMEs and specifically small firms of less than 50 employees show few skills to innovate. It is not so much technological problems which are major but relational aspects, concerning means and even at the level of organizational dynamics. Public policies should integrate these aspects in order to help SMEs to develop new products and processes otherwise than by simple technological help. In other words, they also have to provide support at the strategic level.

12.4 Conclusion

From this inquiry concerning French industrial firms, we were able to verify empirically, on the basis of self-declaration, the Schumpeterian

Table 12.5 Estimation results of 'sale of the innovation'

Competencies	Estimator of $\beta_{T_{PE}}$	Estimator of $\beta_{T_{ME}}$	Estimator of $\beta_{T_{GE}}$	Marginal effect $\beta_{T_{PE}}$	Marginal effect $\beta_{T_{ME}}$	Marginal effect $\beta_{T_{GE}}$	Probabilities
High-intensity sectors							
comp901	−0.56 (−2.00)*	−0.09 (−0.30)	0.36 (1.79)	−0.14	−0.02	0.09	0.50
comp902	−0.26 (−0.94)	0.18 (0.60)	0.94 (4.24)*	−0.06	0.04	0.22	0.61
comp903	0.48 (1.74)	0.56 (1.79)	1.66 (6.08)*	0.09	0.11	0.31	0.75
Medium high-intensity sectors							
comp901	−0.21 (−1.37)	0.11 (0.67)	0.67 (4.09)*	−0.05	0.03	0.17	0.55
comp902	0.05 (0.30)	0.37 (2.17)*	0.80 (4.81)*	0.01	0.09	0.19	0.60
comp903	0.47 (3.02)*	0.64 (3.63)*	1.26 (6.79)*	0.10	0.14	0.27	0.69
Medium weak-intensity sectors							
comp901	−0.65 (−5.78)*	0.14 (1.01)	0.49 (3.76)*	−0.16	0.03	0.12	0.48
comp902	−0.13 (−1.18)	0.19 (1.41)	0.70 (5.21)*	−0.03	0.05	0.17	0.55
comp903	0.26 (2.45)*	0.51 (3.66)*	0.68 (5.09)*	0.06	0.12	0.16	0.61
Weak-intensity sectors							
comp901	−0.90 (−7.86)*	−0.42 (−2.85)*	0.08 (0.45)	−0.21	−0.10	0.02	0.36
comp902	−0.74 (−6.62)*	−0.04 (−0.29)	0.18 (0.98)	−0.18	−0.01	0.04	0.41
comp903	−0.21 (−1.98)*	0.14 (1.00)	0.51 (2.74)*	−0.05	0.04	0.13	0.51

Figures between brackets are the *t*- Student. The values with * are the critical value for 5% non-significant.
Source: SESSI (1997).

conjecture according to which big companies would be more able to implement a research project to innovate. The interest of our research is based on qualitative variables, analysing the first stages of the innovation process, and goes beyond the traditional analyses which were focused on academic indicators such as research expenses or the number of researchers.

Previous studies (Munier, 1999b, 2001) also showed that big companies hold more relational, technical and organizational competencies. This last dimension clearly suggests that the hypothesis of 'Schumpeter Mark I' would be invalidated by the inquiry 'competence to innovate', underlining all the more the advantage of the big company.

Notes

1 Nevertheless the authors clarify that this result is undoubtedly biased since only successful and innovative small firms are taken into account in their sample.
2 Another explanation is based on the theorem of Fisher–Temin (Fisher and Temin, 1973) which shows that an elasticity of R&D/size superior to the unit is a sufficient condition to obtain an elasticity of innovation output/size also superior to the unit (cf. Kohn and Scott (1982) for a determination of the conditions of verification of the Fisher–Temin theorem).

Bibliography

Acs, Z. J. and D. B. Audretsch, *Innovation and Small Firms* (Cambridge, Mass.: MIT Press, 1990).
Acs, Z. J. and D. B. Audretsch, *Innovation and Technological Change: an International Comparison* (Ann Arbor: University of Michigan Press, 1991).
Acs, Z. J. and S. C. Isberg, 'Innovation, Firm Size and Corporate Finance', *Economic Letters*, 35 (1991), 323–6.
Antonelli, C., 'A Failure-inducement Model of Research and Development Expenditure', *Journal of Economic Behavior and Organization*, 12 (1989), 159–80.
Baldwin, W. L. and J. T. Scott, *Market Structure and Technological Change*, A volume in the Economics of Technological Change series edited by F.M. Scherer (London, Paris, New York, Melbourne: Harwood Academic Publishers, 1987).
BETA, *Etude sur les coûts de l'innovation*, Rapport final, contrat de recherche pour le compte du MESR (1995).
Bound, J. C., Z. Cummins, Z. Griliches, B. H. Hall and A. Jaffe, 'Who does R&D and Who Patents?', in Z. Griliches (ed.), *R&D, Patents and Productivity* (Chicago: University of Chicago Press, 1984).
Cohen, W. M., 'Empirical Studies of Innovative Activity', in P. Stoneman (ed.), *Handbook of the Economics of Innovation and Technological Change* (Oxford: Blackwell, 1995).
Cohen, W. M. and S. Klepper, 'A Reprise of Size and R&D', *Economic Journal*, 106 (1996), 925–51.
Cohen, W. M. and D. A. Levinthal, 'Absorptive Capacity: a New Perspective on Learning and Innovation', *Administrative Science Quarterly*, 35 (1990), 128–52.
Cohen, W. M., R. C. Levin and D. C. Mowery, 'Firm Size and R&D Intensity: a Re-examination', *Journal of Industrial Economics*, 35 (1987), 543–63.
Evans, D. and B. Jovanovic, 'Estimates of Model of Entrepreneurial Choice under Liquidity Constraints', *Journal of Political Economy*, 95 (1989), 657–74.
Fazzari, S. M., R. G. Hubbard and B. C. Petersen, 'Financing Constraints and Corporate Investment', *Brookings Papers on Economic Activity*, I (1988), 141–95.
Fisher, F. M. and P. Temin, 'Returns to Scale in Research and Development: What does the Schumpeterian Hypothesis Imply?', *Journal of Political Economy*, 81 (1973), 56–70.
François, J.-P., D. Goux, D. Guellec, I. Kabla and P. Templé, 'Décrire les compétences pour l'innovation. Une proposition d'enquête', in *Innovations et*

Performances (Paris: Editions de l'Ecole des Hautes Etudes en Sciences Sociales, 1999).
Galbraith, J. K., *Le capitalisme américain, le concept de pouvoir compensateur* (Paris: Edition Génin, 1956).
Kamien M. I. and N. L. Schwartz, 'Market Structure and Innovative Activity: a Survey', *Journal of Economic Literature*, 13 (1975), 1–37.
Kamien, M. I. and N. L. Schwartz, *Market Structure and Innovation* (Cambridge: Cambridge University Press, 1982).
Kleinknecht, A., 'Measuring R&D in the Small Firms: How Much are we Missing?', *Journal of Industrial Economics*, XXXVI (2) (1987), 253–6.
Kohn, M. G. and J. T. Scott, 'Scale Economies in Research and Development: the Schumpeterian Hypothesis', *Journal of Industrial Economics*, 30 (1982), 239–49.
Le Bas, C., *Economie du Changement Technique* (Limonest: L'Interdisciplinaire, 1991).
Levratto, N., 'Le financement de l'innovation dans les PME', *Revue d'Economie Industrielle*, 67 (1994), 191–210.
Mairesse, J. and P. Cunéo, 'Recherche–Développement et performances des entreprises: une étude économétrique sur données individuelles', *Revue Economique*, 5 (1985), 1001–41.
Munier, F., 'Taille de la firme et innovation: approches théoriques et empiriques fondées sur le concept de compétence', Thèse de doctorat en Sciences Economiques, Université Louis Pasteur, Strasbourg 1 (1999a).
Munier, F., 'La relation PME–Grande entreprise et compétences pour innover: une vérification empirique sur la base de données individuelles de l'industrie française', *Les 4 Pages des Statistiques Industrielles*, 120 (1999b).
Munier, F., 'Taille de la firme et compétences relationnelles: une vérification empirique sur la base de données individuelles d'entreprises industrielles françaises', *Revue Internationale des PME*, 14(1) (2001), 37–68.
Pavitt, K., M. Robson and J. Townsend, 'The Size Distribution of Innovating Firms in the United Kingdom: 1945–1983', *Journal of Industrial Economics*, 35 (1987), 297–316.
Scherer, F. M., 'Schumpeter and Plausible Capitalism', *Journal of Economic Literature*, XXX (1992), 1416–33.
SESSI, Database 'Compétences pour innover' (Paris: SESSI, 1997).
SESSI, 'Les compétences pour innover' (J.-P. François), 'Chiffres Clés Référence', SESSI (1998a).
SESSI, 'Les compétences pour innover', *Les 4 Pages des Statistiques Industrielles*, 85, SESSI (1998b).
Schumpeter J. A., *Théorie de l'évolution économique* (Paris: Dalloz, 1935). French translation of the second edition of *Theorie der wirtschaftlichen Entwicklung* (Leipzig: Edition Dunker und Humblot, 1926).
Schumpeter, J. A. *Capitalisme, socialisme and démocratie* (Paris: Petite Bibliothéque Payout, No. 55, 1942). French translation of *Capitalism, Socialism and Democracy* (London: G. Allen & Unwin, Ltd, 1942).
Schumpeter, J. A., *Business Cycles*, 2 vols (New York: McGraw-Hill, 1939).
Stiglitz J. and A. Weiss, 'Credit Rationing in Markets with Imperfect Information', *American Economic Review*, 71 (1981), 393–410.

13
The Financing of Innovative Activities by Banking Institutions: Policy Issues and Regulatory Options[1]

Elisa Ughetto

13.1 Introduction

It is a widely held view that firms characterized by high levels of research and development (R&D) spending are very likely to undergo financial constraints. This line of reasoning was originally addressed by two influential papers by Nelson (1959) and Arrow (1962), which pointed to the incomplete appropriability of the returns to R&D as the potential source of the limited private incentives to the allocation of financial resources to basic and applied research. The argument of market failure for R&D investments was later investigated by many researchers in economics and finance (see Hall, 2002 for a review). A common theoretical framework to these studies is that they mostly explain credit rationing or the extension of credit only on unfavourable terms to innovative firms with the presence of information asymmetries between lenders and borrowers. Generally entrepreneurs are better informed than lenders as to the likelihood of success for their innovation projects, and they usually have little incentive to disclose information to investors since this might reveal useful information to competitors (Carpenter and Petersen, 2002; Bhattacharya and Ritter, 1983). Thus investors have more difficulty in distinguishing good projects from bad ones, making credit rationing more probable (Jaffee and Russell, 1976; Stiglitz and Weiss, 1981). Also moral hazard problems can hamper the external financing of innovative investments since entrepreneurs could change *ex-post* their behaviour by replacing low-risk low-return projects with high-risk high-return ones (Hall, 2002; Carpenter and Petersen, 2002).

Alternative or additional explanations of credit rationing highlight that investments in innovation contain a large proportion of intangible assets (predominantly salary payments) which cannot be used as collateral to secure firms' borrowing (Lev, 2001; Bester, 1985). Physical investments designed to embody R&D results are likely to be firm-specific and have little collateral value. Therefore expenditure on R&D can only be backed by the revenue it generates, which is in turn highly uncertain and skewed because R&D projects have a low probability of financial success.

There are in particular two lines of reasoning which have remained relatively unexplored compared to the debate on information asymmetries. The first is related to the unsuitability of the banking system for supporting R&D activities because of a general lack of interest in financing innovation and a shortage of adequate instruments to evaluate innovation projects and innovative firms. What drives a bank to issue a loan is by and large the ability of the obligor to repay the debt. It is a matter of limited interest whether the loan is used to sustain a research activity or the purchase of equipment and machinery. Moreover, the inability of banks to understand and properly classify innovation projects into classes of risk is the result of a lack of specialized technical knowledge. It is also fair to say that, at least until a few years ago, most banks have relied virtually exclusively on subjective analysis to assess the credit risk on corporate loans. The judgement of a banker as to whether or not to grant credit has been mainly based upon considerations on the reputation, leverage and volatility of earnings of the borrower and the presence of collateral (Altman and Saunders, 1998). Qualitative aspects such as investments in intangibles have received very little attention.

An additional cause of the credit-rationing phenomenon is the poor availability of analytical instruments able to capture and correctly estimate the expected future revenues of innovative activities (Encaoua et al., 2000). Investments in intangibles are in most cases not reflected in the balance sheet due to the existence of very restrictive criteria for the recognition of assets and their valuation. As a consequence, financial statements are becoming less informative on the firm's current financial position and future prospects because they do not provide relevant estimates of the value of companies (Cañibano et al., 2000).

These issues, although not sufficiently investigated by scholars, are likely to become relevant in the near future, with the recent display of interest among banks in innovation financing, the adoption of the New Basel Capital Accord framework by banks, and the endorsement of the International Accounting Standards (IAS) by firms.

As far as the first issue is concerned, it is a matter of fact that banks are showing an increased interest in supporting innovation-related activities. This stems from the belief that investments in R&D, information technology and human resources become essential in order to maintain firms' competitive position in a knowledge-based, fast-changing and technology-intensive economy. Banks are moving away from a cautionary approach to innovation financing towards a greater involvement in the support of R&D activities. Indeed, beside government-backed loan schemes both at national and regional level, some specific loan programmes have recently been launched by financial intermediaries.

The second aspect that needs to be underlined concerns the consequences of the implementation of the New Basel Capital Accord on the credit risk assessment of innovative firms. As previously noted, the very core of banking is the classification of loan applications into risk categories, a process which has traditionally been hidden by strict secrecy and complemented by informal practices of 'relationship banking'. Under Basel II banks are prompted to move towards more objectively based evaluation systems and to compete for the best classification procedures. Specifically, they are encouraged to systematically assess risk relative to capital within their organizations, according to an internal ratings-based approach (IRB),[2] subject to the meeting of specific criteria and to validation by the relevant national supervisory authority. This opens up the possibility for banks to use qualitative criteria together with quantitative information in appraising the creditworthiness of their borrowers. A qualitative assessment of a company might take into account the role played by intangible assets[3] as well. Intangibles may encompass patents, trademarks, brands, franchises, R&D, advertising, organizational coherence and flexibility, customer satisfaction, intellectual capital and so forth, depending on the different classification perspectives taken by researchers (see Cañibano et al., 2000 for a review). In other words, the traditional assessment of a borrower's level of risk thought to fit firms whose activity is primarily of a manufacturing or a mercantile nature, could be broadened to reflect intangibles and other qualitative information. Given that intangible assets are likely to be progressively more considered in credit decision making, it follows that firms which would not ordinarily be eligible for bank funding because of limited financial track records and lack of collateral, may have the chance to be granted credit if their qualitative rating is good. In this context innovative firms should theoretically have a lower likelihood of being credit constrained, although this conclusion cannot be taken for granted, at least until the implementation of internal rating models and procedures by major banking institutions is fully completed.

While intangible investments may become an important concern for creditors, the same does not seem to hold for accounting standard-setting bodies. The recently issued IAS regulation embraces a rather restrictive and conservative approach towards accounting for intangibles. IAS 38 considers R&D as a category of internally generated intangible items and as such it requires the full expensing of research, allowing only certain development costs to be carried forward as assets. If most intangible investments are not reflected in the balance sheet but immediately expensed in the income statement, financial statements fail to provide a true estimate of the value of companies. It follows that banks are prevented from getting reliable information on the innovative activity of firms, leading to an unfair or inaccurate credit risk evaluation.

The aim of this chapter is to investigate to what extent the convergence of banks over risk-adjusted capital standards may affect the way in which they screen innovative firms. More precisely, it is worth exploring to what extent banks actually rely on, and will rely on once the Basel II Accord has been implemented, non-financial parameters to assess the creditworthiness of a potential borrower.

Results from a survey conducted in January and February 2006 on a sample of 12 Italian banking groups show that the majority of banks do not consider intangibles as meaningful determinants in credit risk assessment. This could imply that just the implementation of the Accord might not lead to reducing informational asymmetries between lenders and borrowers as might be expected. Hence, if innovative firms show a higher idiosyncratic risk, the bank in its portfolio optimization process might continue either to ask for higher interest rates or simply to deny credit to them. However, such an effect could be compensated by specific measures provided by single financial intermediaries. Current trends suggest that banks are paying increasing attention to the issue of innovation financing, as witnessed by recently launched loan schemes, specifically devoted to sustain technology-based investments.

The remainder of this chapter is organized as follows. Section 13.2 discusses the recent literature on banks' internal rating systems and on the role of non-financial factors in credit risk models. Section 13.3 describes the architecture of the internal rating systems at the banks surveyed. In particular I want to take a closer look at what types of information are being used to determine corporate ratings. Section 13.4 gives an overview of the products banks have in place to finance innovative activities. Section 13.5 offers concluding remarks and some policy indications.

13.2 Overview of the related literature

Since June 2004, when the Basel Committee on Banking Supervision issued a revised framework on International Convergence of Capital Measurement and Capital Standards (hereafter 'Basel II'), the debate on internal ratings has gained increasing importance within banking institutions. Internal rating systems are expected to play a central role not only in credit-granting decisions, but in the determination of regulatory capital adequacy as well (June 2004, para. 6). Whereas academic literature has so far dealt with the various methodologies for the prediction of default and the use of financial ratios in credit risk models, limited interest has been shown in the structure of internal rating systems, in their use of non-financial factors and in the envisaged procedures and internal controls. Earlier empirical analysis of corporate bankruptcy prediction based on financial ratios dates back to Beaver's univariate model (1966). Since then a plethora of multivariate methods have been developed by researchers (see Altman and Saunders, 1998; Szegö and Varetto, 1999 for a review): discriminant analysis, linear probability models, logit and probit regression analysis, or, more recently, recursive partitioning algorithms, multicriteria decision aid methods, expert systems and neural networks. The large number of financial factors proposed in the literature can be gathered into three main groups: those concerning the capital structure, the profitability and the liquidity of a firm. Although accounting based credit-scoring models are widely accepted because of their relatively high discriminatory power, they have been subject to at least three criticisms (see Altman and Saunders, 1998, Szegö and Varetto, 1999). Firstly, they are empirical models lacking an underlying theory of business failure, where explanatory variables are chosen according to their accuracy in predicting default for a specific sample of observations. Secondly, as financial factors are mostly measures that look back in time, these models fail to capture fast-moving changes in borrowers' conditions. Thirdly, these models can hardly maintain their diagnostic potential through time because a variety of elements intervene to jeopardize their temporal stability (for example, structural changes in the economic cycle, inflation rate variations, changes in banking decision-making procedures).

Drawing on the last criticism, researchers have started to include variables other than financial ones in their models, in order to capture macroeconomic, industry-specific and qualitative factors. Macroeconomic variables for failure prediction have been proposed by Foster (1986), Rose et al. (1982) and El Hennawy and Morris (1983). Mensah

(1984) aggregates sample data into four subperiods of the US business cycle from January 1972 to June 1980 (steady growth, recession, steady growth, stagflation and recession) and notes that different economic environments lead to different models for the prediction of failure. Izan (1984) uses industry relative accounting ratios, rather than simple firm-specific accounting ratios, to control for industry variation, and he demonstrates stable classification results both *ex-post* and *ex-ante*. Platt and Platt (1990, 1991) add to their industry-relative model a measure of industry growth to test specific business cycle effects on corporate failure. The industry relative accounting ratio model outperforms the unadjusted model.

Other studies include qualitative data in the analysis of corporate failure. Zopounidis (1987) employs a set of 'strategic criteria' to assess the risk of failure of French enterprises, such as quality of management, R&D level, diversification stage, market trend, market niche/position, cash out method and world market share. Tennyson et al. (1990) consider the information which is contained in annual reports (financial management decisions, influence of external environment on earnings and stockholders, production capacity, variations of exchange rates, new firm strategies), while Laitinen (1993) extends the analysis of the information content of narrative disclosures to their layout, length and language. Keasey and Watson (1987), Daily and Dalton (1994) and D'Aveni (1989) consider qualitative variables related to management characteristics, the composition of the board of directors, corporate governance and a company's reputation.

It is probably fair to say that most of these studies have contributed to shape the architecture of the most recent credit risk systems adopted by financial institutions. As noted earlier, the literature on banks' internal ratings is still scarce, and this is possibly due to the reluctance of banking institutions to disclose information on the structure and input factors of their internal rating systems.

Empirical analysis examines the architecture and use of internal ratings. Udell (1989) looks at the internal rating systems of a sample of Midwestern US banks as part of a broader study of such banks' loan review system. Treacy and Carey (2000) shed light on the use and design of internal risk ratings at large US banks. English and Nelson (1999) describe the internal rating scales of a sample of US banks, reporting the distribution of loans across grades. They also show that ratings are reflected in loan pricing, while non-price terms generally do not rise or fall monotonically with the loan risk rating. An overview of international best practice rating standards in the banking industry is carried out by the Basel Committee Models Task Force (Basel Committee on Banking Supervision, 2000), while information on the operational design of rating systems at Italian banks is provided by Banca d'Italia (2000) and De Laurentis and Saita (2004). Santomero (1997) surveys internal rating systems as a part of a study on banks' credit risk management practices. Other studies use data on internal ratings to perform specific analysis. Machauer and Weber (1998) study loan pricing patterns using German banks' internal ratings. Drawing on bank-internal borrower rating data to evaluate borrower quality, Elsas and Krahnen (1998) provide a direct comparison between housebanks (main banks) and normal banks as to their credit policy in Germany. Grunert et al. (2005) analyse SMEs' credit file data of four major German banks from 1992 to 1996. The authors find evidence that the combined use of financial and non-financial factors leads to a more accurate prediction of future default events than the single use of each of these factors. Brunner et al. (2000) show that 'soft' (qualitative) factors have a significant and positive impact in determining the overall rating of a borrower. Carey (2001) analyses the extent of banks' rating disagreements for given borrowers. Rating disagreements are less likely for large borrowers and for borrowers that have not drawn down much on their lines of credit, while a bit more likely for high-quality borrowers. Tabakis and Vinci (2002) assume that rating inconsistencies derive from a different evaluation of non-financial factors. They therefore compare credit assessments of different financial institutions (rating agencies, banks, other credit assessment institutions). A more normative approach to the issue is taken by Crouhy et al. (2001), who show how an internal rating system can be organized in order to rate creditors systematically. A framework for evaluating the quality of a standard rating system is also suggested by Krahnen and Weber (2001), who advocate 14 principles that ought to be met by good rating practices.

13.3 Results of the survey on internal ratings

In the following sections are presented the results of a survey conducted during January and February 2006 on the 12 main Italian banking groups, selected according to dimensional criteria (with total assets more than €30 billion).[4] Information was collected through extensive structured interviews with bankers operating both in the risk management area and in other units dealing with incentives to R&D and long-term credit. Overall, a total of 24 interviews were conducted, precisely two interviews per banking group.

Table 13.1 Market segmentation and relative weight of business areas by the 12 largest Italian banks (%)

	Large corporate/ corporate	Middle market	Small business
1	29	40	31
2	26	58	16
3	32	30	38
4	28	49	23
5	nd	nd	nd
6	20	40	40
7	45	45	10
8	nd	nd	nd
9	15	45	40
10	20	40	40
11	nd	nd	nd
12	nd	nd	nd

Although the institutions surveyed are the 12 largest banks in Italy, their business segmentation[5] is not the same, as is shown in Table 13.1. In order to maintain the confidentiality of data, banks are numbered randomly, that is being ranked as bank number one does not mean it is the largest Italian bank. Four banks indicated that it was not advisable to release such information. Market segments are defined following Basel II classification (June 2004, paras 232 and 273): large corporate/corporate with total turnover above €50 million: middle market with total annual sales between €5 and 50 million; and small business with turnover below €5 million and exposures to the bank below €1 million. As is evident from Table 13.1, Italian banks are largely oriented towards the middle-market segment. This is not surprising considering that the percentage of SMEs (with less than 500 employees) out of the total number of firms in Italy is greater than 86 per cent.[6] A few differences can be detected when referring to the large corporate/corporate and small business segments. More than half of the respondent banks seem to rely on small business borrowers more than large corporate/corporate customers, whereas the opposite trend is shown by only three banks.

Bankers working in the risk management area were addressed questions regarding the architecture and operating design of their bank's internal rating system (the degree of fineness, the statistical models used, the extent to which judgemental considerations are taken into account, the weight of non-financial factors, the link between ratings and loan price/non-price terms, the organization of the monitoring and rating review activity and so on). Under Basel II it is highly recommended that banks adopt a two-tier rating system that is an independent evaluation of the default probability of the borrower (PD) and of loss given default (LGD), namely the fraction of the loan's value that is likely to be lost in the event of default. While the first dimension is associated with the borrower, regardless of the structure and type of product, the second considers the specific features of the operation, such as its maturity, structure and guarantees.

I decided to maintain the focus of the interviews on the first of these two dimensions, the obligor rating. This was done primarily because few banks have already in place a facility rating that assigns grades to facilities. The overwhelming majority declared that transaction characteristics are explicitly considered in the process of credit risk assessment, but they are still working at developing models of LGD sound enough to get through the validation of the Bank of Italy.

Results are presented in an aggregated form because some banks, by virtue of very strict policies of non-disclosure, explicitly asked me not to make the information released public.

13.3.1 Internal rating systems: architecture

Like a public credit rating produced by agencies such as Moody's or Standard & Poor's, a bank internal rating is meant to summarize the quality of an obligor and the risk of loss due to his failure to repay the debt. While external ratings by agencies have been available for many years, internal ratings by commercial banks began to be introduced only in the last decade.

According to the New Basel Capital Accord, 'the term rating system comprises all the methods, processes, controls and data collection and IT systems that support the assessment of credit risk, the assignment of internal risk ratings and the quantification of default and loss estimates' (Basel Committee, 2004, p. 82).

Of the 12 banks interviewed, all said they had an internal rating system, though a few are currently in an introductory or experimental phase. In particular one bank has just set up the preliminary architecture of the rating system, while at least two banks are testing the soundness of the models and processes with the minimum standards and practice guidelines which have been established by the Basel Committee.

The survey highlighted that internal rating systems differ, at least slightly, across banks in their architecture, methodology and application. The structure of statistical models, the number of grades, the decisions about who assigns ratings or the way in which the review process is conducted, reflect alternative approaches. However, a considerable number of common elements can be identified.

All the banks interviewed base their ratings primarily on a statistical default/credit scoring model. Such models may all be developed internally, as is the case of four banks, in part purchased by suppliers and in part developed internally (five banks) or developed internally with the support of a consultancy firm (three banks). These models are by and large constructed using internal data.

As defined by Brunner et al. (2000), a scoring methodology specifies a number of criteria a_i, one or a number of value functions v_i and an aggregation rule, usually linear, which assigns weights k to single criteria to form an overall score ($v(a)$). The score, which is indicative of a probability of default, is then converted into a rating grade:

$$v(a) = \sum_i k_i v_i(a_i) \tag{13.1}$$

Although this general framework applies to every bank, differences in terms of selected criteria, aggregation rules, weights, rating scales and influence of judgemental factors characterize the several approaches.

Banks reported having several rating models according to customers' segments (for example, large corporate, corporate, SMEs, small business, bank). The number of models goes from an average of 2–3 to about 15. To a considerable extent, such differences may depend on the core business of a bank and on its use of internal ratings for different purposes. Banks in few lines of business are more likely to design their rating system with a limited number of models. As noted earlier, two banks are currently working to improve their rating system and to extend it to more customer segments. It is important to stress, though, that bankers described the models employed in different ways. The low number of models is not always indicative of the degree of accuracy and sophistication of rating systems because macro-models are sometimes divided into several submodels. On average, in each model a further partition can be found, either by sector (for example, real estate, services, industry, commerce), legal form or balance sheet structure (for example, holding, leasing company, manufacturing firm).

Rating models are generally built upon three parts based respectively on financial statement data (cash flow, profitability, short- and long-term debt, debt–equity ratio and so on), behavioural and loss data (both internal and external from Centrale dei Rischi, the national central credit register) and qualitative information. Quantitative criteria are typically backward-looking, while qualitative criteria reflect actual or forward-looking information.

Two-thirds of the banks surveyed used two-stage scoring models which imply that the scores produced respectively by quantitative, qualitative and behavioural models are aggregated by means of a second rule to form the overall score. One bank added to this architecture an additional layer: a market model. Two banks followed the above scheme for corporate and small business segments, while for large corporate borrowers a constrained expert judgement-based process was implemented beside the financial analysis. Only one bank said it had simply a quantitative model and to be about to realize the qualitative part.

The relative importance of each of the above-mentioned modules and the weighting schemes adopted varies widely across banks. Since the data set is not usually homogeneous (it can be that balance sheet information dates back to 5–10 years before, while the qualitative questionnaire has been introduced only one year ahead), banks can use different weighting schemes for quantitative and qualitative data. One bank reported weighting quantitative factors more than qualitative ones mainly for that reason. The vast majority of the banks interviewed outlined that qualitative and behavioural data seem to play a greater role for small business borrowers, where the shortage of financial statement information needs to be counterbalanced somehow. Yet qualitative modules appear to be implemented mainly for small businesses with turnover above €1.5 million.

While in almost every bank qualitative factors enter the statistical model, sometimes they are rather standardized inputs (for example, payment history, industry sector, geographic location). In that case qualitative considerations drive the process of upgrading/downgrading by the rater, who adjusts the rating up or down to a specific limited degree based on his judgement.

There appears to be a relatively limited set of techniques employed in the statistical models. For the vast majority of banks (nine) the calculation engine is based upon logit regressions. To put it briefly, logit analysis uses a set of accounting variables to predict the probability of borrower default which takes a logistic functional form and is constrained to fall between zero and one. Discriminant analysis ranks second, being used by three banks, sometimes together with linear or logistic regressions. Discriminant analysis seeks to find a linear function of accounting variables that best distinguishes between two groups of firms, defaulted and non-defaulted, by maximizing between-group variance while minimizing within-group variance. It is quite surprising that discriminant analysis, which is the most frequently used method in the academic literature dealing with bankruptcy prediction, is so poorly widespread among commercial banks.

Although banks rely on statistical models as important elements of the rating process, expert judgement still plays a fundamental role in assigning a final grade to a counterparty. Especially for large exposures, the current limitations of statistical models[7] are such that processes based on constrained or unconstrained expert judgement are commonly used to deliver a more accurate estimate of risk.

Most of the rating systems were numerical (eight), with the lowest risk borrowers rated 1 and higher ratings implying higher risk. Just one numeric system was in reverse order (1 was the rating for the worst loan rather than the best). Two banks declared having alphanumeric grades (a mixture of letters and numbers), while two others reported following a master scale based on letters similar to that of Standard and Poor's but with a higher granularity in the medium grades. The number of grades conceived by the different banks may vary according to the business segment. The largest part of the banks (eight) surveyed have a standardized number of grades for both corporate and non-corporate borrowers (small and medium enterprises and small business). Retail counterparties are normally rated under a smaller number of classes of risk. Among the banks interviewed, three have a higher grade scale for corporate and large corporate borrowers. This is because for those banking groups which do a significant share of their commercial business in the large corporate and corporate loan market, making fine distinctions among low-risk borrowers is more important in that market than in the middle market. However, it is somewhat difficult to make an accurate taxonomy of the forms of categorization employed by banking institutions because different sorting criteria (for example based on firms' turnover) are used to classify borrowers into business segments. In fact the precise boundary between corporate and middle-market borrowers or between middle-market and small business obligors varies by bank. Larger banks are more likely to have rating systems with a larger and more detailed number of pass categories, though the gap with smaller banks is not so big. Banks with large business loans portfolios (with total assets more than €70 billion) averaged 14 ratings, while those with smaller portfolios (with total assets from €30 to 70 billion) averaged 10.8 for corporate borrowers.

All banks comply with the Basel II requirement (June 2004, para. 404) of having a minimum of seven borrower grades for non-defaulted borrowers and one for those that had defaulted. On average the banks' master scale goes from 9 to 22 non-defaulted categories, with a number of defaulted categories varying from one to four. Only three banks reported conceiving modifiers ('+' or '−') to alpha (two banks) or numeric grades (one bank). Ten banks declared they were satisfied with the actual number of pass grades, while two would like to modify their master scale either by splitting the existing pass categories into a larger number or by adding ± modifiers to the scale in order to reflect a better distribution of exposures across grades. The two banks that expressed the desire to increase the number of grades on their scales have an actual scale of nine classes of risk. Several of the banks' officials indicated that, although internal rating systems with larger number of grades are more costly because of the extra work needed to distinguish finer degrees of risk, they are especially valuable to pricing and capital allocation models. Typically, banks with the highest degree of differentiation appeared to be those using ratings in pricing decisions.

About two-thirds of the banks interviewed declared that the largest part of their corporate loans is concentrated in the upper investment grade categories, revealing a loan distribution skewed towards lower-risk classes. One-third reported a distribution of corporate exposures which approximates a Gaussian distribution, in which loans are centred mostly in the middle classes of risk, while low percentages get into bottom and upper risk grades. A few banks did not answer that question.

The survey asked banks whether there was a direct link between loan terms (such as spreads, size and collateralization) and ratings. More than half of the banks surveyed highlighted that loan pricing can vary depending on the risk rating of the obligor. However, just for three of them pricing always reflects borrower's risk, while for the others ratings are relevant components of pricing decisions although they are not binding. This means that commercial and relationship reasons still play an important role either in the approval process or in the assessment of loan terms. Risk-adjusted pricing is becoming a common practice within large banking groups, while smaller ones are still far away from using ratings to set loan pricing. According to four banks taken from the subsample of the last seven by size, ratings currently influence loan origination and monitoring. The target is to begin to use ratings in pricing, capital allocation models and in setting reserves in the near future.

13.3.2 Internal rating systems: the role of qualitative factors

The survey provides interesting insights into the use of qualitative criteria in credit risk assessment. The results are in line with the requirement of the Basel Committee that banks not only have to consider quantitative but also qualitative factors such as the availability of audited financial statements, the conformity of accounting standards, the depth and skills of management to effectively respond to changing conditions and deploy resources, the firm's position within the industry and its future prospects (June 2004, para. 411; Second Consultative Document, January 2001, para. 265). In all the banks but one (which declared it was about to realize the qualitative part), qualitative inputs, taken from a questionnaire filled in by the line staff, enter the qualitative module of the rating model. All banks reported that the combined use of financial and non-financial factors leads to a more accurate prediction of default events than their single use.

Questionnaires are more or less detailed and extended depending on the bank, but they usually average 20 questions and they are differentiated by sector and borrower. Most of them have been framed internally, while other banks have adopted the CEBI questionnaire, elaborated by Centrale dei Bilanci.

The qualitative analysis is usually concerned with the quality of management, the firm's competitiveness within its industry, as well as the vulnerability of the firm to technological, regulatory and macroeconomic changes. Table 13.2 provides a taxonomy of the main 'soft information' that was cited by bankers as being examined in credit risk assessment.

Since the aim of the study was to explore the extent to which innovation-related parameters are considered in credit ratings, risk managers were asked whether or not they were included in the questionnaire. Nearly two-thirds of the banks surveyed reported having only a few direct questions on patent activity, R&D intensity and innovation capability. However, innovative activity can be inferred from other questions, such as the technological level of facilities or processes, the quality and technological content of goods, the brand, image and reputation of the firm's products. Moreover, the technological capability of a borrower can be further investigated by the relationship manager whenever he is supposed to integrate his own judgemental evaluation to the grade assigned by statistical models.

The reasons why innovation-related parameters do not normally enter statistical models (or have a significantly low weight once entered) mainly relate to two sets of explanations. The first is that it is very difficult for a bank to identify an innovative company, simply because the only reliable information it can get comes from balance sheet data when intangibles are capitalized. However, the decision to capitalize intangible

Table 13.2 Overview of qualitative criteria for credit risk assessment

	Qualitative factors
Business profile	Core business and related business activities Evolutionary stage of activity (start-up, maturity, decline)
Quality of management	Managerial and entrepreneurial capability (flexibility of addressing problems promptly, of introducing or updating methods and technologies when warranted) Risk tolerance and risk propensity Morality (also financial) Professional experience and human resources policies Presence of management succession plans
Ownership structure	Group belonging
Behaviour	Presence of writs, lawsuits or judgements Correct behaviour towards employees
Quality of financial reporting	Clarity, completeness and punctuality in financial data presentation Transparency and prudentiality of accounting information
Industry outlook	Features of the industry and relative position of the firm within its industry Competitive arena and competitive position of the firm
Business risk	Vulnerability to macroeconomic environment (economic downturns, movements in interest rates and exchange rates) Vulnerability to long-term trends that affect demand (lifestyle changes and consumer attitudes) Vulnerability to technological change Impact of environmental and antitrust regulations, fiscal policy, direct and indirect taxation

assets like R&D expenses is in most cases driven much more by fiscal reasons than by disclosure policies. As noted earlier, the implementation of IAS is not going to change anything in this respect.

The second explanation is of a purely statistical nature. Firstly, since the percentage of innovative firms in Italy is very limited, a bank cannot set a default prediction model on the basis of innovative firms' characteristics because a statistical model needs to be as general as possible. Secondly, there are some qualitative components (such as management quality, ownership structure and competitive position) which make the difference, by upgrading or downgrading an obligor rating. Conversely, innovation-related factors are likely to contribute to the final rating not more than a notch. Therefore collecting too many data may not always be helpful.

13.4 Results of the survey on financial support measures to R&D

This section of the chapter gives a brief overview of existing banks' loan schemes devoted to sustain firms' technology-based activities in Italy (see the Appendix for a detailed description of the programmes). As previously anticipated, information was collected from interviews with senior bankers working in the medium–long-term credit divisions.

These consultations indicated that only four banking groups have conceived specific programmes to support R&D investments, with different degrees of specification: Banca Intesa, Sanpaolo IMI, Unicredit and BPU. All the remaining banks said they were participating in government-backed funding programmes, both at national or regional level.[8]

As emerges from Table 13.3, all programmes are devoted to support product and process innovation and other more specific forms of innovation. Technological assessment of the projects is mostly provided by external teams of engineers, except for two banks which have their own internal teams. The loan schemes applied show common features across different banking groups: they are all medium–long-term grants and usually advantageous conditions are applied both in terms of interest rates or collateral requirements. Two banks also provide some consultancy support both prior to the presentation of the project and during its actual implementation.

13.5 Conclusions and policy orientations

It is widely perceived that Italy suffers from an 'equity gap', since the venture capital industry, which should solve the problem of financing innovation for new and young firms, is quite absent. Banks, it is argued, may ration credit to new enterprises, strangling dynamic and innovative future giants at birth. This is because of a lack of track records and collateral and because information about these firms may be limited and asymmetrical, stacked on the side of the borrower at the lender's hazard. Moreover, banks have difficulty in understanding innovative projects since past experience or observed past realizations can offer little guidance in assessing the prospects of truly new projects.

Table 13.3 Banking programmes to sustain innovative activities

	Intesa	Sanpaolo IMI	Unicredit	BPU
Loan schemes	1. IntesaNova	1. Innovation-Buy 2. Applied Research	1. Technological innovation	1. Support to R&D
Eligible projects	• product and process innovation • innovation connected with the diffusion of ICT	• product and process innovation • purchased innovation	• product and process innovation • industrial research	• product and processes innovation • organizational innovation • protection of the environment and energy conservation
Technological assessment	• internal (teams of engineers) • external (network of universities)	• internal (teams of engineers)	• external (national/local associations)	• external (local industrial associations)
Grant decision	• financial/technological evaluation of the project • assessment of the creditworthiness of the borrower (rating between 1 and 6)	• financial/technological evaluation of the project	• financial/technological evaluation of the project	• financial/technological evaluation of the project
Loan terms	• medium-term financing (3–5 years) • no collateral requirement • variable Euribor interest rate + 1–2% range depending on the rating	• medium–long-term financing (3–7 years) • variable Euribor 3 m interest rate • two subsequent anticipations of 50% of the loan • rewards for successful and completed projects	• medium–long-term financing (up to 5 years) • variable Euribor 3 m interest rate, correlated on rating classes • no collateral but covenants	• medium-term financing (up to 5 years) • advantageous conditions
Consultancy support	• prior to the presentation of the project • during the implementation of the project	• prior to the presentation of the project • during the implementation of the project		–

There is recent evidence that this scenario is progressively changing. Banks are encouraged, under Basel II, to incorporate qualitative information in their internal rating models. This is clearly an important issue that cannot be underestimated. Ratings are taking more and more the form of objectively based 'screening devices' that can alleviate asymmetric information problems between borrowers and lenders, and in doing so they account for information other than simply financial to appraise the creditworthiness of obligors. In that way innovative firms should theoretically have the chance of being less credit constrained.

However, the evidence suggests that innovation-related parameters are not yet taken into account by Italian banks in a systematic way. In fact the majority of banks do not consider intangibles as meaningful determinants in credit risk assessment. This is primarily the result of a regulatory caveat which prevents banking institutions from inferring appropriate information on firms' innovative activity from financial statements, rather than banks' reluctance in considering such factors to a greater extent.

Even though a wider recognition of qualitative elements in credit risk assessment is on the way, just the implementation of the Accord might not lead to reducing informational asymmetries between lenders and borrowers, at least in the short run. This seems to be acknowledged by the fact that banks have started to conceive some forms of credit support for R&D activities which would not be necessary if the implementation of the Basel II Accord could really lead banks to screening innovative firms in a better way.

Given these current trends, I positively advocate a renewed role for the banking system in supporting science and technology-based activities. As a matter of fact, the expansion of banks' activities in terms of innovation financing is likely to have a positive and strong impact on the whole Italian industrial system, largely constituted by small and medium enterprises. Banks are territorially distributed and may respond efficiently to SMEs, strongly featured locally and mostly incapable of building lasting relationships with international capital. Therefore the banking system could bring about the innovation-based development process of the Italian industrial system, helping it to reach that dimensional threshold to get to other forms of financing.

Indeed, working on the criticalities which have traditionally characterized borrower–lender relationships is a necessary requirement if banks intend to start offering to their customers not only products, but also solutions. In this respect universities and research centres may contribute to alleviating information asymmetries, by giving a technology assessment of innovation projects and collecting all the relevant information to orientate credit-granting decisions.

In conclusion, the future challenge for economic development is to plan the emergence of virtual spaces of overlapping institutional spheres for science and technology-based activities. A new organizational environment should emerge in which industry, financial institutions, universities/research centres and government tend to integrate their own interests and goals when carrying out, financing and regulating investments in R&D.

13.6 Appendix

I give a brief overview of the products developed by the banks interviewed to sustain R&D-intensive activities.

Intesa Group

Intesa Group has launched two specific programmes related to R&D support: IntesaNova and Eurodesk.

IntesaNova

IntesaNova is a funding scheme purposely designed for companies involved in substantial research activities. Firms can submit their research project to the bank and get financing on advantageous conditions and without collateral requirements. Innovation projects above €200,000 up to €1 million are normally assessed by an internal team of engineers. For higher levels of complexity or cost amounts above €1 million, the bank gets the support of a network of outstanding Italian universities (Politecnico di Torino, Politecnico di Milano, Università degli Studi di Trento Politecnico di Bari). The evaluation of the project implies an assessment of its costs, degree of innovation, realization time, as well as considerations on the competitive position of the firm and its implementation capacity. On the basis of a technological/financial evaluation of the project and the creditworthiness of the firm (which can be eligible only if it has a rating ranging between 1 and 6), the bank issues a medium-term loan (3–5 years), with a variable Euribor interest rate plus a 1 or 2 percentage range depending on the rating. Universities also provide technological support when the project reaches the implementation phase (auditing of the product/process development, prototype realization, laboratory experimentation, consultancy for patenting, marketing of technologies). The programme currently applies to two product families: product and process innovation and innovation connected with the diffusion of information and communication technologies. In only one year of effectiveness of the programme, about 800 projects have been examined and 600 financed.

Eurodesk

In the light of the 7th Framework Programme of the European Commission, Banca Intesa intends to support the participation of Italian companies through cooperation with research centres and universities. This means offering consultancy for the entire life of approved projects and acting as a *trait d'union* with the academic world. In this perspective IntesaNova can be extended, thanks to EU funding opportunities, to a wider spectrum of R&D activities and universities involved.

Sanpaolo IMI Group

Sanpaolo IMI Group has recently launched two programmes specifically devoted to support R&D and technologically driven investments: Applied Research and Innovation-Buy. These schemes are thought to respond to companies' requirements about demand and supply of innovation. Firms willing either to develop a technologically advanced product, service or process, or to buy innovation from external sources, can submit their project to the bank which, upon acceptance, will finance it on favourable terms. A technological evaluation of the project is carried out by an inside team of engineers, specializing in different technological sectors. Marketing and profitability analysis complement the technological validation of the project.

Applied Research

Applied Research is aimed at financing R&D projects directed either at the realization/completion of new technologically advanced products, processes or services or to the technological improvement of existing products, processes or services. It is a long-term loan scheme with a loan period between three and five years, including a pre-amortization that ends six months after the end of the project. The loan covers up to 100 per cent of the cost of the project, which does not have to be below €250,000 or above €4,000,000. The project can last one or two years. A variable interest rate Euribor 3m (3 months) applies for the entire loan period. An interesting point that needs to be underlined in this respect is that the bank anticipates 50 per cent of the loan when the contract is drawn up and another 50 per cent when half of the cost of the project is overcome. Moreover for completed and successful projects a kind of reward is applied: a 20 per cent spread reduction if the project is brought to an end and a two-year increase of amortization if it has a positive outcome. One year after the launch of the programme at the end of 2004, 550 projects had been financed at a total cost of €550 million. Around 30 projects were not accepted. The large majority of the funded projects are devoted to product innovation (65 per cent). A smaller percentage applies to process innovation (21 per cent) and product/process innovation (14 per cent). Request of funds is markedly affected by geographical location: firms from the north of Italy (Lombardy and Piedmont above all) have been granted more funds, although a notably reverse trend is shown by the region of Campania. Innovation projects mainly concern the mechanical and ICT sectors.

Innovation-Buy

Innovation-Buy is aimed at financing the purchase of innovation in its different forms (technologies, tangible and intangible goods, training). It is a medium–long-term loan scheme with a loan period between five and seven years, including a pre-amortization of two years. The loan covers up to 100 per cent of the cost of the purchase, which does not have to be below €250,000 or above €4,000,000. The investment can last up to 18 months. A variable interest rate Euribor 3 m applies for the entire loan period. Even in this case the bank anticipates 50 per cent of the loan when the contract is drawn up and another 50 per cent when half of the cost of the investment is overcome. For completed projects a kind of reward is applied: a 15 per cent spread reduction for five-year transactions and 10 per cent spread reduction for transactions beyond five years. The programme was started in November 2005. A pilot experiment took place in December 2005 in the Brescia area and within four weeks 34 demands were presented for a total financing of €30 million.

Unicredit Group

Technological innovation

Unicredit provides medium–long-term loans (up to five years) to sustain firms in their product and process innovation and industrial research. Interest is calculated on Euribor 3m and it is correlated with rating classes. A technological evaluation of the project is carried out by national or local associations. Covenants but not collateral are required.

BPU

Support to R&D

BPU has recently created a credit line to sustain R&D activities. A technology check-up of companies' research projects is carried out by local industrial associations. Upon such evaluation, BPU issues a medium-term loan (up to five years), including a pre-amortization of 12 months. The loan covers up to 100 per cent of the cost of the project. Projects can be devoted to the realization of new products or processes, to technological and organizational innovation, to the protection of the environment and energy conservation. The amount of loans, which is about to be extended, is around €70 million.

Notes

1 I am deeply indebted to Luigi Buzzacchi, Mario Calderini and Giuseppe Scellato for their helpful insights on previous versions of the chapter. I also thank Steven Fazzari and Bruce Petersen for fruitful discussions on this topic and participants at the International Symposium 'Knowledge, Finance and Innovation 2006' for useful comments and suggestions.
2 The IRB approach gives the bank varying degrees of autonomy in the estimate of the parameters determining risk weightings and consequently, capital requirements: under the Foundation only the probability of default (PD) is internally estimated, while under the Advanced a bank can also produce its own estimates for the loss given default (LGD) and exposure at default (EAD).
3 IAS 38 defines intangible assets as non-physical and non-monetary sources of probable future economic profits accruing to the firm as a result of past events or transactions.
4 The banking groups are the following: Intesa, Unicredit, Sanpaolo IMI, Capitalia, Monte dei Paschi di Siena (MPS), Banca Nazionale del Lavoro (BNL), Banche Popolari Unite (BPU), Banco Popolare di Verona e Novara, Banca Antoniana Popolare Veneta, Banca Popolare dell'Emilia Romagna, Bipiemme, Banca Lombarda e Piemontese.
5 I just consider the percentage of claims on corporate, SMEs, small business segments of the total claims on firms. Loans to sovereign entities, banks and retail are therefore excluded.
6 Istat, I gruppi di imprese in Italia, 2003.
7 It is indeed very difficult to distinguish between defaulted and non-defaulted firms for large corporate customers which are usually characterized by low default rates and consequently to construct a statistical model. Therefore judgemental factors tend to have a more prominent role in corporate and large corporate lending rather than in middle-market or small business lending.
8 Capitalia, through MCC (Mediocredito Centrale), is responsible for the management of numerous national subsidy programmes devoted to the support of R&D activities (FAR, FIT, Fondo agevolazione regionale, Fondo Capitale di Rischio, Fondo Garanzia). Surveyed banks reported being involved in subsidy lending for different ones of these government-backed loan schemes.

Bibliography

ABI, *Albo dei gruppi bancari: struttura, movimentazione e dati di bilancio* (Rome: ABI, 2005).

Altman, E. I. and A. Saunders, 'Credit Risk Measurement: Developments over the Last 20 Years', *Journal of Banking and Finance*, 21 (1998), 1721–42.

Arrow, K., 'Economic Welfare and the Allocation of Resources for Invention', in R. Nelson (ed.), *The Rate and Direction of Incentive Activity: Economic and Social Factors* (Princeton: Princeton University Press, 1962).

Banca d'Italia, 'Modelli per la gestione del rischio di credito. I "ratings" interni', *Tematiche istituzionali*, April (2000).

Basel Committee on Banking Supervision, 'Range of Practice in Banks' Internal Rating Systems', Discussion paper series, January (2000).

Basel Committee on Banking Supervision, *International Convergence of Capital Measurement and Capital Standards*, Bank of International Settlements (2004).

Beaver, W., 'Financial Ratios as Predictors of Failure', *Journal of Accounting Research*, 4 (1966), 71–102.

Bester, H., 'Screening vs. Rationing in Credit Markets with Imperfect Information', *American Economic Review*, 75 (1985), 850–5.

Bhattacharya, S. and J. R. Ritter, 'Innovation and Communication: Signaling with Partial Disclosure', *The Review of Economic Studies*, 50 (1983), 331–46.

Brunner, A., J. P. Krahnen and M. Weber, 'Information Production in Credit Relationships: on the Role of Internal Ratings in Commercial Banking', Working paper 2000/10, Centre for Financial Studies, Frankfurt am Main (2000).

Cañibano, L., M. García-Ayuso and P. Sanchez, 'Accounting for Intangibles: a Literature Review', *Journal of Accounting Literature*, 19 (2000), 102–30.

Carey, M., 'Some Evidence on the Consistency of Banks' Internal Credit Ratings', Working paper, Federal Reserve Board (2001).

Carpenter, R. and B. Petersen, 'Capital Market Imperfections, High-tech Investment and New Equity Financing', *Economic Journal*, 112 (2002), 54–72.

Crouhy, M., D. Galai and R. Mark, 'Prototype Risk Rating System', *Journal of Banking and Finance*, 25 (2001), 47–95.

Daily, C. and D. Dalton, 'Corporate Governance and the Bankruptcy Firm', *Strategic Management Journal*, 37 (6) (1994), 1603–17.

D'Aveni, R., 'Dependability and Organizational Bankruptcy: an Application of Agency and Prospect Theory', *Management Science*, 35 (9) (1989), 1120–38.

De Laurentis, G. and F. Saita, *Rating interni e controllo del rischio di credito: Esperienze, problemi, soluzioni* (Rome: Bancaria Editrice, 2004).

El Hennawy, R. and R. Morris, 'The Significance of Base Year in Developing Failure Prediction Models', *Journal of Business Finance and Accounting*, 10 (1983), 209–23.

Elsas, R. and J. P Krahnen, 'Is Relationship Lending Special? Evidence from Credit-file Data in Germany', *Journal of Banking and Finance*, 22 (1998), 1283–316.

Encaoua, D., F. Laisney, B. H. Hall and J. Mairesse, *Economics and Econometrics of Innovation* (Amsterdam: Kluwer, 2000).

English, W. B. and W. R. Nelson, 'Bank Risk Rating of Business Loans', in *Proceedings of the 35th Annual Conference on Bank Structure and Competition*, May (1999).

Foster, G., *Financial Statements Analysis* (London: Prentice-Hall, 1986).

Grunert, J., L. Norden and M. Weber, 'The Role of Non-financial Factors in Internal Credit Ratings', *Journal of Banking and Finance*, 29 (2005), 509–31.
Hall, B., 'The Financing of Research and Development', *Oxford Review of Economic Policy*, 18 (1) (2002), 35–51.
Izan, H. Y., 'Corporate Distress in Australia', *Journal of Banking and Finance*, 8 (1984), 303–20.
Jaffee, D. and T. Russell, 'Imperfect Information, Uncertainty and Credit Rationing', *The Quarterly Journal of Economics*, 90 (1976), 651–66.
Keasey, K. and R. Watson, 'Non Financial Symptoms and the Prediction of Small Company Failure', *Journal of Business Finance and Accounting*, 14 (1987), 353–5.
Krahnen, J. P. and M. Weber, 'Generally Accepted Rating Principles: a Primer', *Journal of Banking and Finance*, 25 (2001), 3–23.
Laitinen, E., 'The Use of Information Contained in Annual Reports and Prediction of Small Business Failure', *International Review of Financial Analysis*, 3 (1993), 55–72.
Lev, B., 'Intangibles: Management, Measurement and Reporting', *Brookings Institution Papers* (2001).
Machauer, A. and M. Weber, 'Bank Behavior Based on Internal Credit Ratings of Borrowers', *Journal of Banking and Finance*, 22 (1998), 1355–83.
Mensah, Y., 'An Examination of the Stationarity of Multivariate Bankruptcy Prediction Models: a Methodological Study', *Journal of Accounting Research*, 22 (1) (1984), 380–95.
Nelson, R., 'The Simple Economics of Basic Scientific Research', *Journal of Political Economy*, 49 (1959), 297–306.
Platt, H. D. and M. B. Platt, 'Development of a Class of Stable Predictive Variables: the Case of Bankruptcy Prediction', *Journal of Business Finance and Accounting*, 17 (1) (1990), 31–51.
Platt, H. D. and M. B. Platt, 'A Note on the Use of Industry-relative Ratios in Bankruptcy Prediction', *Journal of Banking and Finance*, 15 (1991), 1183–94.
Rose, P. S., W. T. Andrews and G. A. Giroux, 'Predicting Business Failure: a Macroeconomic Perspective', *Journal of Accounting and Finance*, 6 (1) (1982), 20–31.
Santomero, A. M., 'Commercial Bank Risk Management: an Analysis of the Process', *Journal of Financial Services Research*, 12 (1997), 83–115.
Stiglitz, J. and A. Weiss, 'Credit Rationing in Markets with Imperfect Information', *American Economic Review*, 71 (1981), 393–410.
Szegö, G. and F. Varetto, *Il rischio creditizio. Misura e controllo* (Turin: Utet, 1999).
Tabakis, E. and A. Vinci, 'Analysing and Combining Multiple Credit Assessments of Financial Institutions', Working paper, European Central Bank, 123 (2002).
Tennyson, B. N., R. W. Ingram and M. T. Dugan, 'Assessing the Information Content of Narrative Disclosures in Explaining Bankruptcy', *Journal of Business Finance and Accounting*, 17 (3) (1990), 391–410.
Treacy, W. F. and M. Carey, 'Credit Risk Rating Systems at Large US Banks', *Journal of Banking and Finance*, 24 (2000), 167–201.
Udell, G. F., 'Loan Quality, Commercial Loan Review, and Loan Officer Contracting', *Journal of Banking and Finance*, 13 (1989), 367–82.
Zopounidis, C., 'A Multicriteria Decision Making Methodology for the Evaluation of the Risk of Failure and an Application', *Foundations of Control Engineering*, 12 (1) (1987), 45–67.

14
Innovation and the Profitability Imperative: Consequences on the Formation of the Firm's Knowledge Capital

Blandine Laperche

14.1 Introduction: the firm's objectives and the profitability imperative

14.1.1 The firm's objective: a theoretical debate

In neoclassical economics, the firm is seen as a 'black box', i.e. an entity receiving flows of raw materials and turning out flows of processed or finished products. The purpose of such an entity is to maximize its profit, being limited only by its resources. However, since the 1960s, this restrictive vision of the firm's goals has been largely questioned and has been commonly replaced by a more complex spectrum of objectives.

Due to the number and the economic weight of large corporations, they can hardly be considered as an exception to the pure and perfect competition rule. If we accept that the large corporation is generated by the market economy, our understanding of concentration, and thus of the large corporation, recalls in certain ways the approaches of K. Marx, J. A. Schumpeter, J. Robinson (in her writings of the 1960s and 1970s, see Laperche, 2001a) and J. K. Galbraith (Laperche, 2006).

For Galbraith (1967), the power exercised over firms and society shifted from capital to organized competence, i.e. 'the technostructure' in the twentieth century, as it has shifted from land property to capital in the nineteenth century. The transfer of power from capital to the technostructure goes together with the revelation of the new goals of the planning system given impetus by the technostructure members' new individual motivations, other than profit maximizing or remuneration.

The profit-maximizing assumption was applicable when capital was the prevailing production factor, as the capitalist, who had full decision-making power in the firm, would maximize his contribution, i.e. money. The situation has changed with the technostructure, as it contributes specific skills and organization capacity. According to this logic, what the technostructure maximizes is no longer money, but its skills and organizing capacity.

The technostructure members and the planning system therefore have other objectives. Profit-making is essential to ensure the technostructure's self-sufficiency and survival, but it is no longer the only objective. Other goals in line with the aspirations of the members of the technostructure and with the objectives of the social body are more and more important, as the growing size of sales which ensures the growth of the technostructure (high-responsibility jobs, promotion, protection against contraction) and technical virtuosity which is a sign of prestige. Likewise, the individual motivation system also broadens out to integrate identification with the technostructure and the wish to adapt in addition to financial reward.

Moreover, as the power is controlled by the technostructure, the objectives of the planning system and the members of the technostructure will generate internal efficiency made even stronger by the discipline imposed by collective decision-making. As a matter of fact, the power of the managers quickly alarmed economic observers: those managers can take advantage of the information they have on daily business to make decisions that are opposed to the interests of the shareholders and all the stakeholders of the firm: employees, customers and suppliers. Already in the 1930s, their personal enrichment was a cause of concern. However, according to Galbraith, in the technostructure, as the decision-making process is collective, the pursuit of individual interest becomes highly improbable. Such discipline also counteracts the profit maximization assumption which is the only one accepted by traditional theory to explain the motivations of the firms. As a matter of fact, the managers who have the power in the firm do not receive the benefits themselves. If the assumption was maintained, all the members of the technostructure would be regarded as maximizing the profits of the firm not for themselves but for remote stockholders.

If this analysis (variety of objectives and motivation in the big firm) correctly describes the reality of large corporations' management in the 1960s and 1970s, is it still the case at the beginning of the twenty-first century?

14.1.2 Globalization and the profitability imperative

Since the beginning of the 1980s, three main changes have modified the strategy and the organization of big companies, and hence have modified the hierarchy of their objectives.

First of all innovation, i.e. new combinations (Schumpeter, 1942), has become the main differentiation strategy in global competition (compared to prices in the Fordist mode of production). From an organizational perspective, this evolution has led to a transformation of the firm from a hierarchical to a networked organization. A large modern corporation can be sketched as a network of units owned by a central firm (usually a holding company) and the other kinds of activities linked by contract (partnerships, subcontracting, licensing) (Andreff, 2003; Chesnais, 1994; Uzunidis et al., 1997). The firm focuses on its core activities (the ones that will reinforce its innovation capacity) which it owns and achieves the production and marketing of goods mainly through contracts with other independent entities (subcontractors, licensees, etc.). The same flexibility exists for the management of employees: the firm focuses on a stable core of managers in R&D, financial and administrative departments. It uses more diverse forms of work and contracts of employment (in terms of working time, salaries, place of work, job content) to manage the other employees. The first main change is thus that innovating has become a fundamental objective of the firm, imposed by competition. This idea is commonly shared by business theories (Porter, 1990; Tidd et al., 2005; Uzunidis, 2004).

Networked firms have gained greater flexibility, thus enabling them to adjust to the evolution of demand. With this greater internal flexibility are associated increased options in the ways firms manage their assets at international level (external flexibility). The globalization of corporate strategies refers to their liberty or flexibility in the management of human, financial, scientific and technical assets at international level. Networked firms are organized at a global level, benefiting from the competitive advantages of potential host territories. Holding companies are located in areas with low or even non-existing taxation. Research and development laboratories are set up in areas where financial, scientific and technical resources are abundant. Production plants select attractive countries in terms of specialization and labour costs as well as transport infrastructures. Goods are marketed in all financially solvent areas worldwide. This globalization of firms' strategies has been made possible by liberal policies of market (goods and services, labour, financial) integration, developed and diffused through international organizations (WTO, IMF and World Bank) (Michie, 2003, Milward, 2003). Internal and external flexibility have an important impact on the functioning of the working groups: the worldwide dispersion of the members of the technostructure makes it more difficult for them to identify with the specific objectives of the organization (as for example innovation, growth of sales, of size). The organic solidarity (in Durkheim's words, see Durkheim, 1930 [1996 edn]) that ties the members of working teams is closely related to physical vicinity, which information technologies reproduce only very imperfectly. Financial motivation would thus come back as the main motivation in the big global firm. If identification and the wish to adapt are present, the pecuniary motivation however represents the universal objective of the members of global working teams, all the more as the evolutive character of networks makes employees more vulnerable to the possible strategic changes decided by big multinational firms (naturally, the employees who are far from the decision-making centres are hit first, but executives are also more frequently hit by reorganizations) under the pressure of institutional stockholders. The increase of opportunist behaviours at the end of the 1990s and the beginning of the 2000s (which J. Galbraith and J. Sawyer relate in Chapters 1 and 3) can thus be understood as the result of this undermined cohesion within the large corporation (Stiglitz, 2003; Dietrich and Sharma, 2006).

The third change is as a matter of fact the increasing role of finance in the management of companies (Chesnais, 2004; Plihon, 2002). The different steps of financial market deregulation and liberalization have produced an interconnected global market. New types of investors (pension funds, insurance companies, investment funds) are investing in big enterprises worldwide. Due to their main activity (e.g. managing employees' pension funds), they feel less concerned by the development of such companies than by the amount of the dividends to be received. Their fluctuating behaviour (they 'vote with their feet'), dependent on the level of the price earning ratio, has important implications in the management of such corporations. In particular, the objective of profit maximization, linked to the increase of the shareholder's value, comes back as one of the most important.

The 'profitability imperative' is the result of this new context. It means that in order to keep the precious new institutional investors, managers of big globalized corporations have to boost shareholder value. The increase in shareholder value will moreover be profitable to them, as they have often become, thanks to stock options plans, shareholders of the companies they manage. This profitability imperative is, however, in contradiction with the necessity to boost innovation, which needs long-term and uncertain investments (as developed by J. Courvisanos in Chapter 6). The remainder of this chapter precisely deals with the consequences of the contradiction between those two objectives on the way big firms manage the innovation process. We thus study the strategies implemented by firms (notably big multinationals in the industrial sector) to constitute and protect their 'knowledge capital'. 'Knowledge capital' is here defined as a combination of technological knowledge and information produced, acquired and systematized by one or several firms to be used in the process of value creation. In the second part of the chapter, we define and explain the formation of the firm's knowledge capital. External means of formation of a firm's knowledge capital are now of growing importance (through cooperative agreements with other firms – big or small – or with universities and research labs and also through informal contacts, hiring short-term researchers, licensing, etc.). However, in-house strategies are still essential, especially to integrate the new scientific and technical information into the enterprise knowledge stock. This growing importance of external means of formation of knowledge capital is according to us explained by today's profitability imperative and its related necessity to reduce the risk, cost and length of technical progress.

However, if firms rely more and more on external relations to constitute their own knowledge capital, how do they protect it from competition? In the third part of the chapter, after recalling the main protection tools used by enterprises (intellectual property rights, secrecy, standards) we present the idea according to which the recent trends to extend patenting possibilities to new fields (information technology, genetics), closer to the scientific border and to harmonize intellectual property rights at the global level are driven by the same profitability imperative. As firms are more and more open to their environment, they need to have a wider and stronger protection of their own knowledge base in which patents have a major part to play. Moreover, new legal practices – notably patent pools – secure the development of collective relations. In the same time, they officialize the oligopolistic appropriation of knowledge capital. What are the consequences of this growing contradiction between, on the one hand, the socialization of knowledge capital and, on the other, the tendency toward its growing oligopolistic appropriation? The final remarks open the debate on some of these consequences, notably on the danger hanging over the scientific commons and on the place of knowledge and universities in modern societies.

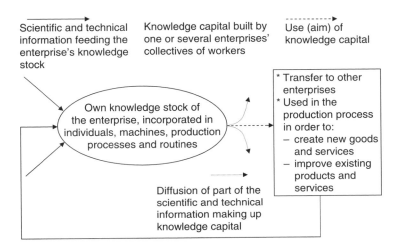

Figure 14.1 Knowledge capital

14.2 The firm's knowledge capital

14.2.1 Knowledge capital: definition, role and formation

We can define knowledge capital as the set of scientific and technical knowledge and information produced, acquired, combined and systematized by one or several firms for productive purposes. Knowledge capital (see Figure 14.1) refers to the accumulated knowledge of one or several linked firms (embedded in the individuals – know-how – machines, technologies and routines of the enterprise) which is continuously enriched by information flows and which is used in the production process or more globally in the value creation process. Thus, it is a dynamic concept – a process – that defines the knowledge accumulated by one or several firms and continuously enriched and combined in different ways, and eventually used or commercialized. This productive aim – the creation of value – is the main characteristic which turns knowledge into 'capital'.

A firm may use its knowledge capital in a value creation process by:

1. Simply selling this knowledge base to another enterprise (e.g. the selling of a computer program). Thus, knowledge capital (embodied in the software) is transferred to another enterprise which can use it in its production process;
2. Using this knowledge capital in its own production process. In this case, knowledge capital can be considered as a means to produce or to

improve goods and services and as a tool for reducing its production process completion time.

Theoretically, the notion of knowledge capital is based on the definitions and/or on the economic developments of three key concepts/notions: knowledge, firm and capital. The economic analysis of knowledge has changed over time. Neoclassical economists first considered technical progress as exogenous and knowledge as a public good, notably characterized by its non-appropriability (Arrow, 1962; Nelson, 1959). As for technical progress, the firm has also long been considered as a 'black box', and before the contributions of the 1930s (Berle and Means, 1932; Chamberlin, 1933; Robinson, 1933), the firm did not very much catch the contemporary economists' attention (except for the major contributions of A. Marshall and T. Veblen).

After the Second World War, the firm became a *complete* object of study, place of production of new knowledge (Penrose, 1959) and symbol of modern capitalism (Galbraith, 1967). The interest which arose from the work of Schumpeter (1950, 1983) on the role of innovation in economic dynamism but also from Solow's work (1957) on the residual technical progress and economic growth gave birth to new analysis aiming at explaining the origin of innovation and of knowledge. The evolutionist theory and more globally the resource-based approaches stress the learning processes which are at the origin of the firm's own technological trajectories and put forward the double nature of knowledge, codified and tacit, which makes its appropriability possible (almost in part) (Dosi et al., 2000; Nelson and Winter, 1982; Nonaka and Takeuchi, 1995). The new growth theories have taken account of those new developments and also show the role of the state in the allocation and appropriation of fundamental resources for growth (Aghion and Howitt, 1998). The most recent developments insist on the role of external knowledge, and notably on the way firms capture external information in their environment (which explains the key role of proximity and interactions) and transform it into the enterprise's own knowledge (Antonelli, 2005). Knowledge and innovation are thus considered as collective processes.

The notion of knowledge capital is built on these main evolutions in the analysis of knowledge and the firm. But it does not neglect the crucial contributions of early authors. The developments of the classical economists already stressed the collective nature of production and of innovation. A. Smith (1776) considered the technical and social division of labour as a means to increase the productive and innovative power of labour. J.-B. Say (1815) analysed the links between the scientist, the entrepreneur and the worker. K. Marx (1857) showed how production is based on the combined workforce – appropriated by capital – of the collective of workers. The notion of knowledge capital also borrows from the classical economists their dynamic conception of capital. In this approach, capital is not only a stock of resources available for production; it is a process which indicates the constant renewal and the productive use of this stock.

To sum up, knowledge capital is fed by modern or more ancient approaches of knowledge creation, coordination and diffusion. It is centred on firms' capabilities, with the analysis of innovation being often related to the resource-based approaches which have become the mainstream perspective in the business management discipline (Courvisanos, 2006). Knowledge capital also aims to integrate the value creation process (which, for instance, can take the form of the production and diffusion of a new machine). This determines the integration of information in the knowledge stock, the combination of information and knowledge, the codification of tacit knowledge and the diffusion of knowledge. In brief, this is the constitution of knowledge capital.[1] With this particular focus on the aim – the value creation process – we also reintegrate in the analysis the tensions, linked to the relations of power existing between firms of different size and strength and which particularly occur in the current context of constitution and protection of knowledge capital.

The formation of a firm's knowledge capital has old roots in an economic system characterized by change (Laperche, 1998). At the beginning of industrial capitalism, states were the main investors in the production and protection of knowledge, with a view to strengthening their political and economic power. Throughout the twentieth century, when the national systems of innovation (Freeman, 1987; Lundvall, 1992) were better structured, the states were the first investors in R&D, conducted by public or private institutions. Their role has also consisted in promoting an attractive and incentive economic context (including notably systems to protect intellectual property). Nowadays, in all industrialized countries, corporate investment in gross domestic expenditures (GERD) largely exceeds public investment. At the beginning of the 2000s, enterprises account for more than 60 per cent of the OECD's GERD (OECD, 2003).

The formation of the enterprise's knowledge capital implies the gathering of different types of inputs, i.e. human resources (researchers, engineers), tangible resources (machines, tools) and intangible ones (patents, software, information). The enterprise has to produce and appropriate scientific and technical knowledge in order to expand the

Table 14.1 Means of formation of the firm's knowledge capital

In-house means	External means
Investment in human resources Investment in and management of R&D and means of production (tangible and intangible)	Contracts with other firms (including licensing) Contracts with institutions: e.g. university research labs (including licensing and hiring of short-term researchers) More informal contacts

knowledge base it has already accumulated. Different means are used by the enterprise, which we can call for one part in-house means (investment and management of human resources, R&D and tangible and intangible resources), and for the other part external means (contractualization with firms and other institutions and more informal contacts) (see Table 14.1).

14.2.2 Socialization of the formation of the firm's knowledge capital

Echoing the important strategic and organizational changes of big firms (see introduction), important transformations can be presented in the way firms form their knowledge capital.

Given the strong competitive context based on technological performance and 'permanent innovation' (Foray, 2004), enterprises try to develop their knowledge base; they focus on their core activities and try to develop their competencies within these activities. Moreover, they are more and more interested not only in applied research but in basic research as well, which is one of the explanations of the interactive character of the innovation process (Kline and Rosenberg, 1986).

The second new trend is the globalization of R&D. If R&D has long been regarded as a case of non-globalization (Patel and Pavitt, 1991), the studies conducted in the 1990s show that this trend is gaining ground, whatever the focus of specific studies: in terms of foreign-based laboratories (Bartlett et al., 1997; Florida, 1997; Madeuf et al., 1997; Madeuf and Lefebvre, 2002); in terms of patents and technological flows (OECD, 2003); and in terms of R&D partnerships (Archibigu and Iammarino, 2002; Hagedoorn, 2002). The 2005 issue of the *World Investment Report* also reports a sharp increase in R&D globalization, which, however, mostly concerns the most applied parts of the R&D process (UNCTAD, 2005).

Finally, firms are developing their knowledge base more and more thanks to the network(s) in which they are involved. The issue of networks is now considered as a challenge in the economics of innovation, which starts 'from the recognition that innovation and industry are highly affected by the interaction of heterogeneous actors with different knowledge, competencies and specialization, with relationships that may range from competitive to cooperative, from formal to informal, from market to non market' (Malerba, 2006, p. 15). The link between innovation and networks which has been made theoretically and empirically (Pyka and Küppers, 2002), shows that after relying on states' investments and on their own competencies, big firms are relying more and more on external means of formation of knowledge capital. The increase in R&D partnerships (to which we have already referred) is a first illustration of that trend. Another support for this idea is the role played by innovative start-ups and their links with bigger enterprises. When they are not taken over by bigger firms, start-ups are often linked by contract with them and the former usually take part in their funding (through corporate ventures) (Laperche and Bellais, 2001). In this strategy, the large corporation does not bear alone the risk inherent in the development of new technology, and shares it with its partner (here, the start-up). This strategy has been largely used during the 1990s by big American firms to enrich their knowledge base (Tidd et al., 2005, pp. 425–63). Outsourcing is another way to reduce the cost of technological development (and hence a strategy to increase profitability) and also results in the expansion of networks. It is very often used in the software industry (software design outsourced to Bangalore, India, for example). A last, but crucial, example is the closer relationship between universities and enterprises. This type of relationship has been allowed by law in the US since the early 1980s (notably the Bayh Dole Act) – and was further adopted by many countries – thus facilitating the signature of contracts as well as technology transfers between enterprises and universities and more informal contacts between enterprise staff and scholars (Jaffe, 2000; Mowery et al., 2001; Laperche, 2002).

The purpose of all these strategies is to reduce the cost, risk and length of technical progress and hence increase the short-term return on investment in scientific and technical fields. This purpose is all the more important as the complexity of technological development increases, which implies a collective process able to innovate more quickly and with fewer risks. Due to the profitability imperative, the big enterprise develops external means of formation of the knowledge base, which are both less risky and less costly. This does not mean, however, that the firm no longer makes investments in-house, as this kind of investment is crucial to understanding and absorbing the scientific and technical development achieved by other institutions on their own base (Rosenberg, 1990; Cohen and Levinthal, 1990). This trend shows that the formation of knowledge capital is socialized, i.e. several institutions (big or small enterprises, research laboratories, etc.) take part in the formation of one's firm knowledge capital. However, in this context, how do firms cope with the question of the protection of knowledge capital? Does this trend towards socialization mean that the need and the willingness for appropriation are on the wane?

14.3 The appropriation of knowledge capital

14.3.1 Patents, secrecy and technical standards: a portfolio of protection tools

Intellectual property rights are the first tools that come to mind when dealing with the protection of an enterprise's knowledge capital. As a matter of fact, the knowledge capital of an enterprise is composed of information and knowledge – embedded in individual minds but also in its machines, products, industrial processes, software programs – produced, acquired and managed by the enterprise itself.

Regarding technical invention, industrial property rights, and notably patents, are the most important tools of protection[2] (Foray, 2004; Levêque and Menière, 2003). As they grant a 20-year temporary monopoly (under the TRIPs agreement[3]), patents represent for firms an incentive to invest in the design and production of inventions. Thanks to patents, inventors can secure and increase their return on investment (by working their inventions, or giving this possibility to other firms through licence agreements). They also have the possibility to protect themselves from possible patent infringement. Moreover, thanks to their patents, inventors can find partners or financial resources more easily.

However, patents are not always positively judged by enterprises: they are regarded as expensive, a fact which explains that big enterprises are the main patent holders. As a matter of fact, the cost of a patent not only refers to the cost of obtaining it but also includes the cost of possible lawsuits, which increases in a global context, due to the different appropriability regimes existing worldwide. Moreover, not everything can be patented. Patents only protect technical inventions, thus excluding substantial portions of the knowledge capital of the enterprise (however, the concept of technical invention is today ambiguous, considering the extended capacity to patent in new fields, see below). Finally, patents diffuse too much information. As a counterpart of the temporary monopoly given to the inventor, patent laws include the obligation to disclose the scientific and technical information. The aim is to ensure the cumulative nature of technical progress (Scotchmer, 1991). Patents are thus not always perfect tools to protect inventions, and many recent studies have shown that the part of secrecy and lead time has increased during the 1990s as a protection mechanism for both product and process innovation (Levin et al., 1987; Harabi, 1997; Cohen et al., 2000). Two questions arise from these results: is it preferable to protect invention through patents or secrecy? How can a firm build a 'lead time' over its competitors?

Secrecy is another tool used by enterprises to protect their knowledge capital, perhaps the oldest one. In their rules, guilds laid down the necessity to keep their production methods secret. A. Smith explained in the *Wealth of Nations* (1776) that trade secrets were at the origin of extraordinary profits in capital (1976 edn, p. 87). K. Marx also exposed in *Capital* (1867) that the first manufactures employed 'some almost idiot workers' for operations containing trade secrets (1982 edn, I, p. 350). However, a secret is difficult to keep, all the more so in a context of global competition. Secrecy cannot provide absolute protection, because some process characteristics can be identified through an analysis of the final product (reverse engineering, espionage, spillovers). Moreover, Schumpeter (1950) has shown that innovation is the result of the collective work of a bureaucracy, the 'technostructure', in the words of Galbraith, and thus indiscretions or the trading of trade secrets (Von Hippel, 1988) can reduce the interest of secrets, all the more so when turnover is high (as is often the case in clusters and science parks, Carnoy et al., 1997). And, finally, secrets have a cost. In enterprises which traditionally use secrets to protect their products or processes, a very specific organization of work is implemented, characterized by a very developed division of all tasks so that very few of the highest managers know the entire production process (it is for instance the case for Michelin, in France, also Coca-Cola in the USA). In fact, industrial secrets and patents can be associated. Secrecy can be used for the protection of process innovation and patents for products. If the market is new and competition not too harsh, if the research is not completed, if the financial resources of the enterprise are not sufficient, keeping the invention secret can be a first step, before filing a patent.

The best knowledge capital protection consists in continuously being ahead of one's competitors. In order to build this leadership over their competitors, firms try to erect entry barriers using formal or informal strategies. Informal strategies include the accumulation of tacit knowledge (Polanyi, 1958), which needs time and learning and is difficult to evaluate and transfer, precisely because of the knowledge tacitness and stickiness. The accumulation of tacit knowledge thus enables the firm to build its own specific advantage (Nelson and Winter, 1982; Dosi, 1988). Formal strategies implemented to build a lead time rely on a mix of patents and standards, held by several players linked by strategic partnerships. Strategic R&D partnerships aim at sharing the costs and the risk of the development of new technology and, concurrently, at constructing entry barriers. According to Foray (1990), standards are normally built by consensus, taking account of the interests of all stakeholders, but in sectors where technological progress is fast, standards are often defined collectively by major firms, anticipating the consensus and the birth of the relevant innovations. Standards are defined without any consensus between producers and users, and technologies are imposed on all the other firms in the sector. Thanks to the patents they hold on the standardized techniques, and their licensing strategies (patent pooling), major firms accumulate technological rents that can be reinvested in new technological development and thus build lead time over their competitors (Laperche, 2001b). However, the definition of a collective standard or the domination of one standard is not an easy process and is the theatre of strategies and tactics (Besen and Farrell, 1994), including standard wars which are very common in information technology sectors (Shapiro and Varian, 1998).

To summarize, firms – individually and collectively – use a portfolio of tools to protect their knowledge capital, arbitrating between the advantages and restrictions of different kinds of protection mechanisms and continuously trying to increase their leadership over their competitors. However, in order to assess properly the protection strategies of the firm's knowledge capital, we have to take account of today's characteristics of its elaboration. External means of formation of knowledge capital now have a growing importance – due to what we called the profitability imperative and its related necessity to reduce the risk, cost and length of technical progress.

14.3.2 Towards an oligopolistic appropriation of knowledge capital

The strategies implemented by firms to protect their knowledge capital have also changed. Two main elements can be presented: the first one is the extension of patentability to new scientific and technical fields, which can be considered as an answer to the greater interest of firms in innovation and research (both applied and basic); the second element is the emergence of new possibilities in terms of collective protection of knowledge capital. These two elements show the growing contradiction between, on the one hand, the socialization of the knowledge capital and, on the other, the tendency towards its oligopolistic appropriation.

As firms are more and more open to their environment, they need to provide their own knowledge base with wider and stronger protection. The recent trend towards extending patentability to new fields and closer to the scientific border can be regarded as an answer to this growing need for protection (the global protection given by the TRIPs agreement also favours their appropriation strategies) (Gallini, 2002; *Revue d'économie industrielle*, 2002; Laperche, 2004). Back in the 1980s, in a context of decreasing competitiveness and serious challenge by Japanese enterprises, the US made substantial changes in the IPR, and notably in the fields of biotechnologies and information and communication technologies (ICT), i.e. the embryonic technologies of the time. Software programs were traditionally protected through copyright (this was explained by the fact that, as they are composed of mathematical algorithms, they were excluded from patentability, just like natural laws, scientific theories, natural phenomena, abstract ideas, formulae and methods). However in the US, a case law led to the patentability of computer programs (*Diamond v. Diehr*, 1981). Computer program patentability ensued from the explanation that a computer program represents an invention (in terms of process) and from the fact that it produces a useful, concrete and tangible result. The patentability of computer programs paved the way for the possibility to patent business models (*Street Bank v. Signature*, 1988). In Europe, even though the legal context is not clear, many software patents have been granted. The origin of the extension of patentability to living organisms can also be found in the US, and was based on the argument that a living being produced by a non-natural process (apart from human beings) is eligible for patent. Then patentability was extended to recombinant DNA (1980), to transgenic animals (the 'onconmouse patent' in 1988) and to human gene and research tools (DNA sequences). In Europe, a 1998 directive specified the patentability of genes and of partial gene sequences.

Moreover, the scope of industrial property rights was widened at the end of the 1990s, with the Trade Related Industrial Property Rights (TRIPs). This agreement allows patentability in all technological fields and harmonizes the protection period covered by patents – 20 years. This agreement is managed by the WIPO and WTO, and any infringement of this agreement can lead to commercial sanctions. Thus, it creates a favourable context for the global diffusion of patented technology (Maskus and Reichman, 2004). All of these institutional changes evidence a greater need for protection, requested by firms themselves. This greater appropriation need can be linked to what we have called the profitability imperative. Global corporations have to innovate in order to be competitive. The complexity, but also the rapid pace, of technological progress ('permanent innovation') leads to increase in the cost and hence in the risk of the innovation process, which has nonetheless to be reduced if firms want to keep their precious investors. To reduce the cost, risk and length of the innovation process, firms rely on their own capabilities but also on the resources offered by their networks. However, being more open to their environment, they become more vulnerable, all the more so when appropriability regimes are different in the countries they are active in. That is why corporate lobbying is a major explanatory element of that legal evolution, as reported by S. K. Sell in the case of the TRIPs agreement (Sell, 2003).

The strong increase in the number of patents (filed and granted), notably in the US system, has given rise to debates in the US but also in Europe, where the US model is traditionally regarded as a source of inspiration. The extension of patentability to new subject matters and closer to the scientific border has been seen as a source of concern because it may block the exploitation of knowledge and it is at the origin of litigation costs (Gallini, 2002). As a result, even if short-term returns may grow, the increase in transaction costs may hinder innovation in the long run and thus may be counterproductive and contradictory to the interests of industry in the long run (reduction of the stock of 'free' knowledge from which the firm may draw; in other words, reduction of externalities stemming from investment in R&D).

Some legal solutions are proposed to reconcile the incentives to innovate and the dissemination of knowledge, such as compulsory licensing, non-exclusive licences, modifying the duration and the breadth of patents (O'Donoghue et al., 1998; Scotchmer, 2004). But another type of solution to these restrictions has been found in the way firms manage their industrial property rights. Some studies have shown that building patent pools could be a solution to the blocking of knowledge or could prevent litigation (Clark et al., 2000; Shapiro, 2001; Choi, 2003).

A patent pool can be defined as 'an agreement between two or more patent owners to license one or more of their patents to another or third party', or more precisely as 'the aggregation of intellectual property rights which are the subject of cross-licensing, whether they are transferred directly by patentee to licensee or through some medium, such as a joint venture, set up specifically to administer the patent pool' (Clark et al., 2000, p. 4). Patent pooling is not new, but during most of the twentieth century in the US, this practice was regarded as illegal under antitrust laws. However, since the beginning of the 1980s, discussions have gained ground on the positive impacts of patent pooling, and led to the *Antitrust Guidelines for the Licensing of Intellectual Property* in 1995 (issued by the US Department of Justice and the Federal Trade Commission) which recognizes that 'patent pools can have significant pro-competitive effects' (Clark et al., 2000, p. 6). According to this guideline, an intellectual property policy is pro-competitive when it

- integrates complementary technologies
- reduces transaction costs
- clears blocking positions
- avoids costly infringement litigation
- promotes the dissemination of knowledge.

The same report states that the benefits of such a strategy are the elimination of problems caused by blocking patents, the increase in the disclosure of information between patent pool members, the reduction of licensing transaction costs and the distribution of risk: 'Like an insurance policy, a patent pool can provide incentive to further innovation by enabling its members to share the risks associated with research and development. The pooling of patents can increase the likelihood that a company will recover some, if not all, of its costs of research and development efforts' (Clark et al., 2000, p. 9). The latter argument shows that the patent pooling strategy, which is gaining ground in new technology sectors (like ICT and biotechnology), is driven by the same profitability imperative which also explained the development of external means of formation of knowledge capital. It also supports the idea of a growing private and oligopolistic appropriation of knowledge capital which is quite contradictory to the socialization of its formation. In other words, even if the formation of knowledge capital depends on interdependent relations between increasing numbers of institutions (big firms, small concerns, research labs, etc.), only a few firms are able to appropriate the returns on their investment. Moreover, patent pooling can encourage the development of monopolistic behaviours (such as high prices, imposition of 'invalid' technologies, and so on).

14.4 Final remarks: impacts of the contradiction

What are the implications of this growing contradiction between, on the one hand, the socialization of knowledge capital formation (importance of external sources of knowledge) and, on the other, the tendency towards its growing oligopolistic appropriation. We can put forward two main issues arising from this contradiction, which we regard as crucial.

The first implication, which is a much studied subject, is the danger hanging over the scientific commons. This evolution is not new. K. Marx, notably in *Grundrisse* (1857), explained that within capitalism, when industry is well developed, invention becomes 'a branch of business' and thus science becomes a productive factor of capital (Uzunidis, 2003). However, during a long period it did not catch the economists' attention. Today, many of them put forward the idea according to which the closer relations between firms and universities or public labs and the new capacity to patent nearer the scientific border result in the reduction of autonomous science, which is nevertheless crucial for the development of science, technology and economic progress in general. 'While the privatization of the scientific commons has been relatively limited so far, there are real dangers that, unless halted, soon important portions of future knowledge will be private property and fall outside the public domain, and that could be bad for both the future progress of science, and for technical progress' (Nelson, 2004, p. 455).

Another issue is that of the social role of knowledge development and thus, the place of universities in modern societies. There are not many things in common between the university aiming at developing knowledge without practical goals and today's 'entrepreneurial universities' (Etzkowitz, 2003). If universities are regarded as competitors for innovative enterprises, the question of the blocking of future research arises again. This is suggested by a recent decision of the US Federal Circuit explaining that, as conducting research work, basic or applied, is part of the core business of a university, it is quite reasonable under the law for a patent holder to require that the university take out a licence before using patented material in research. 'After this ruling it is highly likely that patent holders will act more aggressively when they believe that university researchers may be infringing on their patents' (Nelson, 2004, p. 466). If there is no blocking in the case the university can afford the cost of the licence, we can imagine that it would generate or accentuate hierarchy between universities, between those which can conduct research work, and thus will attract the best academics and the best students, and the others. This greater privatization raises the issue of the right balance between public and private activities, already raised for example by John Kenneth Galbraith in the *Affluent Society* (1958), in which he explained that the disequilibrium between the two kinds of activities led to 'private opulence and public squalor'. This means that all non-competitive activities in the short run could be neglected by private enterprises, which could have negative impacts on the future of our societies.

Notes

1 Note that the question of the difference between knowledge and information is not settled, as the symposium published in the review *Economic Journal Watch* (2 (1), 2005) demonstrates. According to us, the difference between information and knowledge can be studied as a difference of level in the categories of knowledge but also as a different method of accounting. In the first case, the whole of knowledge appears as 'Russian dolls': it is defined as a set of knowledge bits more or less systematized and the knowledge bits are defined as sets of information. In the second case, knowledge is considered as a stock and information as a flow. Knowledge and information are thus linked: information flows which are integrated in the enterprise have a founding power over the accumulated knowledge. The information flows organize the accumulated knowledge in a definite aim: create a new product for example. Knowledge and information are the result of a working process: knowledge is a result of theoretical and practical work aimed at improving the understanding of natural and social facts. Information describes and diffuses the knowledge produced and needs a work of selection of the most pertinent bits of knowledge. This means that information is the diffused result of a knowledge bit. Not all knowledge will become information, whether because the knowledge has still not reached a sufficient level of formalization in order to give a better understanding of natural and social facts (knowledge is still a set of hypotheses), or because this individual or collective knowledge is without immediate usefulness in an aim of – for example – merchant commercialization.
2 However, in fact, firms often use different titles of industrial property rights for one invention: brand or trademark for the name of their products, and registered design for the form of their products. Firms can also resort to copyright.
3 The Agreement on Trade-Related Aspects of Intellectual Property Rights (TRIPS), which is annex 1C of the Marrakesh Agreement establishing the WTO, was signed in Marrakesh, Morocco, on 15 April 1994.

References

Aghion, P. and P. Howitt, *Endogeous Growth Theory* (Cambridge, Mass.: MIT Press, 1998).
Andreff, W., *Les multinationales globales* (Paris: La Découverte, 2003).

Antonelli, C., 'Models of Knowledge and Systems of Governance', *Journal of Institutional Economics*, 1 (2005), 51–73.
Archibugi, D. and S. Iammarino, 'The Globalisation of Technological Innovation: Definition and Evidence', *Review of International Political Economy*, 9 (2002), 98–122.
Arrow, K. J., 'Economics Welfare and the Allocation of Resources for Invention', in R. R. Nelson (ed.), *The Rate and Direction of Inventive Activity: Economic and Social Factors* (Princeton, NJ: Princeton University Press of NBER, 1962), pp. 609–25.
Bartlett, C. A., Y. Doz and G. Hedlund (eds), *Managing the Global Firm* (London: Routledge, 1990).
Berle, A. A. and C. G. Means, *The Modern Corporation and the Private Property* (New York: Macmillan, 1932).
Besen, S. and J. Farrell, 'Choosing how to Compete: Strategies and Tactics in Standardization', *Journal of Economic Perspectives*, 8 (1994), 117–31.
Carnoy, M., M. Castells and C. Benner, 'Les marchés de l'emploi et les pratiques en matière d'emploi à l'ère de la flexibilité: étude de cas de la Silicon Valley', *Revue internationale du travail*, 136 (1997), 29–54.
Chamberlin, E. H., *The Theory of Monopolistic Competition* (Cambridge, Mass.: Harvard University Press, 1933).
Chesnais, F., *La mondialisation du capital* (Paris: Syros, 1994).
Chesnais, F. (ed.), *La finance mondialisée* (Paris: La découverte, 2004).
Choi, J. P., 'Patent Pools and Cross-Licensing in the Shadow of Patent Litigation' (Michigan State University, http://www.msu.edu, 2003).
Clark, J., J. Piccolo, B. Stanton and K. Tyson, *Patent Pools: a Solution to the Problem of Access in Biotech Patents?* (USPTO, http://www.uspto.gov, 2000).
Cohen, W. and D. Levinthal, 'Absorptive Capacity: a New Perspective on Learning and Innovation', *Administrative Science Quarterly*, 35 (1990), 128–52.
Cohen, W. M., R. R. Nelson and J. P. Walsh, 'Protecting their Intellectual Assets: Appropriability Conditions and why US Manufacturing Firms Patent (or Not)?' (National Bureau of Economic Research Working Paper 7552, 2000).
Courvisanos, J., 'Galbraith and the Political Economy of Technological Innovation: Critical Perspectives and a Heterodox Synthesis', in B. Laperche, J. K. Galbraith and D. Uzunidis (eds), *Innovation, Evolution and Economic Change. New Ideas in the Tradition of Galbraith* (Cheltenham: Edward Elgar Publishing, 2006), pp. 205–28.
Dietrich, M. and A. Sharma, 'The Corrupt Corporation: a Galbraith-inspired Analysis', in B. Laperche, J.K. Galbraith and D. Uzunidis (eds), *Innovation, Evolution and Economic Change. New Ideas in the Tradition of Galbraith* (Cheltenham: Edward Elgar Publishing, 2006), pp. 162–84.
Dosi, G., 'Sources, Procedures and Microeconomic Effects of Innovation', *Journal of Economic Literature*, 26 (1988), 1120–71.
Dosi, G., R. R. Nelson and S. Winter, *The Nature and Dynamics of Organizational Capabilities* (Oxford: Oxford University Press, 2000).
Durkheim, E., *De la division du travail social* (1930) (Paris: PUF, 1996).
Etzkowitz, H., 'Research Groups as 'Quasi-Firms': the Invention of the Entrepreneurial University', *Research Policy*, 32 (2003), 109–21.
Florida, R., 'The Globalization of R&D: Result of a Survey of Foreign Affiliated Laboratories in the USA', *Research Policy*, 26 (1997), 85–103.

Foray, D., 'Exploitation des externalités de réseau versus évolution des normes', *Revue d'économie industrielle*, No. 51 (1990), 113–40.

Foray, D., *The Economics of Knowledge* (Cambridge, Mass.: The MIT Press, 2004).

Freeman, C., *Technology and Economic Performance: Lessons from Japan* (London: Pinter Publishers, 1987).

Galbraith, J. K., *The Affluent Society* (Boston: Houghton Mifflin Company, 1958).

Galbraith, J. K., *The New Industrial State* (Boston: Houghton Mifflin Company, 1967).

Gallini, N. T., 'The Economics of Patents: Lessons from Recent US Patent Reform', *Journal of Economic Perspectives*, 16 (2002), 131–54.

Hagedoorn, J., 'Inter-Firm R&D Partnerships: an Overview of Major Trends and Patterns since 1960', *Research Policy*, 31 (2002), 477–92.

Harabi, N., 'Les facteurs déterminants de la R&D', *Revue Française de Gestion*, 114 (1997), 39–51.

Jaffe, B., 'The US Patent System in Transition: Policy Innovation and the Innovation Process', *Research Policy*, 29 (2000), 531–7.

Kline, S. J. and N. Rosenberg, 'An Overview of Innovation, National Academy of Engineering', *The Positive Sum Strategy: Harnessing Technology for Economic Growth* (Washington, DC: The National Academy Press, 1986).

Laperche, B., *La firme et l'information* (Paris: L'Harmattan, 1998).

Laperche, B., 'Les ressorts du monopole, essai sur l'hérésie de Joan Robinson', *Innovations, Cahiers d'économie de l'Innovation*, 14 (2001a), 33–54.

Laperche, B., 'Brevets et normes techniques. De l'incitation à l'invention au contrôle de l'innovation', in B. Laperche (ed.), *Propriété industrielle et innovation* (Paris, L'Harmattan, 2001b), pp. 81–98.

Laperche, B., 'The Four Key Factors for Commercialising Research', *Higher Education and Management Policy*, OECD, 14 (2002), 149–75.

Laperche, B., 'Patentability: Questions about the Control of Strategic Technology', in C. A. Shoniregun, I. P. Chochliouros, B. Laperche, O. Logvynovskiy and A. Spiliopoulou-Chochliourou, *Questioning the Boundary Issues of Internet Security* (London: E. Centre for Economics, 2004), pp. 117–40.

Laperche, B., 'Large Corporations and Technostructures in Competition', in B. Laperche, J. Galbraith and D. Uzunidis (eds), *Innovation, Evolution and Economic Change: New Ideas in the Tradition of Galbraith* (Cheltenham: Edward Elgar Publishing, 2006), pp. 142–61.

Laperche, B. and R. Bellais, 'Entrepreneurs, capital-risque et croissance des grandes entreprises', *Problèmes économiques*, 2704–2705 (2001), 14–21.

Lévêque, F. and Y. Menière, *Economie de la propriété industrielle* (Paris: La découverte, 2003).

Levin, R., R. Klevorick, R. Nelson and S. Winter, 'Appropriating the Returns from Industrial Research and Development: a Review of Evidence', *Brookings Papers on Economic Activity*, 3 (1987), 783–831.

Lundvall, B. A., *National Systems of Innovation: Towards a Theory of Innovation and Interactive Learning* (London: Pinter Publishers, 1992).

Madeuf, B. and G. Lefebvre, 'Innovation mondiale et recherche localisée. Stratégies "technoglobales" des groupes. Le cas français', *Innovations, Cahiers d'économie de l'innovation*, 16 (2002), 9–27.

Madeuf, B., G. Lefebvre and A. Savoy, 'De l'internationalisation à la globalisation de la R&D industrielle: l'exemple de la France', *Innovations, Cahiers d'économie de l'innovation*, 5 (1997), 55–92.
Malerba, F., 'Innovation and the Evolution of Industry', *Journal of Evolutionary Economics*, 16 (2006), 3–23.
Marx, K., *Grundrisse* (1857), vol. II (Paris: Anthropos, 1977).
Marx, K., *Le Capital* (1867) (Moscow: Editions du Progrès, 1982).
Maskus, K. E. and J. H. Reichman, 'The Globalization of Private Knowledge Goods and the Privatization of Global Public Goods', *Journal of International Economic Law*, 7(2) (2004), 279–320.
Michie, J. (ed.), *The Handbook of Globalisation* (Cheltenham: Edward Elgar Publishing, 2003).
Milward, B. *Globalization? Internationalization and Monopoly Capitalism* (Cheltenham: Edward Elgar Publishing, 2003).
Mowery, D. C., R. Nelson, B. N. Sampat and A. Ziedonis, 'The Growth of Patenting and Licensing by US Universities: an Assessment of the Effects of the Bayh Dole Act of 1980', *Research Policy*, 30 (2001), 99–119.
Nelson, R. R. 'The Simple Economics of Basic Scientific Research', *Journal of Political Economy*, 67 (1959), 297–306.
Nelson, R. R., 'The Market Economy, and the Scientific Commons', *Research Policy*, 33 (2004), 455–71.
Nelson R. R. and S. G. Winter, *An Evolutionary Theory of Economic Change* (Cambridge: Harvard University Press, 1982).
Nonaka, I. and H. Takeuchi, *The Knowledge Creating Company* (New York: Oxford University Press, 1995).
O'Donoghue, T., S. Scotchmer and J.-F. Thisse, 'Patent Breadth, Patent Life and the Pace of Technological Progress', *Journal of Economics and Management Strategy*, 2 (1) (1998), 1–32.
OECD, *Science, Technology and Industry Scoreboard 2003* (Paris: OECD, 2003).
Patel, P. and K. Pavitt, 'Large Firms in the Production of the World's Technology: an Important Case of "Non Globalisation"', *Journal of International Business Studies*, First Quarter, 22 (1) (1991), 1–21.
Penrose, E.T., *The Theory of the Growth of the Firm* (Oxford: Blackwell, 1959).
Plihon, D., *Le nouveau capitalisme* (Paris: Dominos, Flammarion, 2002).
Polanyi, M., *Personal Knowledge, Towards a Post Critical Philosophy* (London: Routledge and Kegan Paul, 1958).
Porter, M. E., *The Competitive Advantage of Nations* (London: Macmillan, 1990).
Pyka, A. and G. Küppers, *Innovation Networks: Theory and Practice* (Cheltenham: Edward Elgar Publishing, 2002).
Research Policy, 'The Internationalization of the Industrial R&D', Special Issue, 28 (2, 3), (1999).
Revue d'économie industrielle, 'Les droits de propriété intellectuelle, nouveaux domaines, nouveaux enjeux', Special Issue, 99 (2002).
Robinson, J., *The Economics of Imperfect Competition* (London: Macmillan, 1933).
Rosenberg, N., 'Why do Firms do Basic Research (with Their Own Money)?', *Research Policy*, 19 (1990), 165–74.
Say, J.-B., *Catéchisme d'économie politique*, in *Cours d'économie politique et autres essais* (1815) (Paris: GF Flammarion, 1996).

Schumpeter, J. A., *Capitalism, Socialism and Democracy* (1942) (New York: Harper and Row, 1950).
Schumpeter, J. A., *The Theory of Economic Development* (1911) (New Brunswick and London: Transaction Publishers, 1983).
Scotchmer, S., 'Standing on the Shoulders of Giants: Cumulative Research and the Patent Law', *Journal of Economic Perspective*, 5 (1991), 29–41.
Scotchmer, S., *Innovation and Incentives* (Cambridge: MIT Press, 2004).
Sell, S. K., *Private Power, Public Law. The Globalization of Intellectual Property Rights* (Cambridge: Cambridge University Press, 2003).
Shapiro, C., 'Navigating the Patent Thicket: Cross Licensing Patent Pools, and Standard Setting', in A. Jaffe, J. Lerner and S. Stern (eds), *Innovation Policy and the Economy*, vol. I (Cambridge: MIT Press, 2001).
Shapiro, C. and H. Varian, *Information Rules: a Strategic Guide to the Network Economy* (Boston, Mass.: Harvard Business School Press, 1998).
Smith, A., *An Inquiry into the Nature and Causes of the Wealth of Nations* (1776) (Paris: Coll. Folio essais, Gallimard, 1976).
Solow, R., 'Technical Change and the Aggregate Production Function', *Review of Economics and Statistics*, 39 (1957), 313–20.
Stiglitz, J. E., *The Roaring Nineties* (New York: W.W. Norton, 2003).
Tidd, J., J. Bessant and K. Pavitt, *Managing Innovation. Integrating Technological, Market and Organizational Change* (Chichester: J. Wiley & Sons, Ltd, 2005).
UNCTAD, *World Investment Report 'Transnational Corporations and the Internationalization of R&D'* (New York and Geneva: United Nations, 2005).
USPTO, *Trilateral Statistical Report* (various editions: 1996, 2000, 2003), http://www.uspto.gov
Uzunidis, D., 'Les facteurs actuels qui font de la science une force productive du capital. Le quatrième moment de l'organisation de la production', *Innovations, Cahiers d'économie de l'Innovation*, 17 (2003), 51–78.
Uzunidis, D., *L'innovation et l'économie contemporaine* (Brussels: De Boeck, 2004).
Uzunidis, D., S. Boutillier and B. Laperche, *Le travail bradé* (Paris: Economie et Innovation, L'Harmattan, 1997).
Von Hippel, E., 'Trading Trade Secrets', *Technology Review*, 61 (1988), 58–64.

Index

accounting
  generally accepted principles, 52
  industry relative accounting ratio, 229
  market-to-market, 52
accounting ratios, 229
accumulation foundations, 27
Adelphia, 51
aggregate demand, 35, 41
  and investment, 43–4
alertness in action, 158
alternative technology options, 191–2, 201–2
Anglo-Saxon tradition of entrepreneurship, 149
Antitrust Guidelines for the Licensing of Intellectual Property, 263
Apollo, 169
Apple, 169
Applied Research, 243–4
appropriation
  of knowledge capital, 258–63
areas of specialization, 163
Armstrong, Edwin, 95
Asset Manager Code of Professional Conduct, 54
Association of Chartered Financial Analysts/Centre for Financial Market Integrity, 54
Austrian tradition of entrepreneurship, 149, 157–60

banks, 224–47
  internal rating systems, 232–6
  market segmentation, 231
  programmes to sustain innovation, 240
  see also finance
Basel Capital Accord I, 8, 103
Basel Capital Accord II, 225, 226
Bayh Dole Act, 257
Beaver's univariate model, 228
big box retailers, 55

'big finance', 1
big firms, 67, 69–75, 77–8
  financial means, 210
  R&D costs, 211
  technological intensity, 215
biotechnology, 76, 103, 111, 123, 261
Bogle, John, *The Battle for the Soul of Capitalism*, 53
Boulding, K., 162
BPU, 244–5
Breaking the Short-Term Cycle, 54
Bretton Woods Agreements, 75, 101, 114
Breusch–Pagan test, 176, 179
budget constraints, 41
business development, 90
business profile, 238
business risk, *see* risk
Business Roundtable/Institute for Corporate Ethics, 54
Business Unit logic, 179–80

capacity utilization, 42
capital, xi
  entrepreneurial, 194
  financial, 112–16
  human, 194, 198
  knowledge, *see* knowledge capital
  physical, 194, 198
  socialization of, 51
capital acquisition costs, 198–9
capital investment, 190–1
capitalism, 20–5, 68
  crisis and renewal, 22–5
  foundations of innovation, 20–2
  post-industrial, ix
  professionals', ix–x, xiii
  restructuring, 112–16
  totalitarian, 110
  transnational, 110
capitalist efficiency, 121–3
capital stock, 46
cartels, 94

centralization, 21
centres of excellence, 31, 90
  see also universities
cephalisation, 153
chief executive officers, 16
China, 16–17
Cisco Systems, 16
closed innovation systems, 90
closed systems, 172
Coase, Ronald H., 150, 154–7
Coca-Cola, 75, 259
commercialization of knowledge, 117
Commission on Industrial
    Productivity, 56
commitments, 42
commodification, 116
Commodore, 169
communications technology, 261
competencies, 207–23
  complex, 212
  elementary, 212
  financial, 212, 214, 217–18, 219
  R&D, 217, 218
  sale of innovation, 220, 221
competition, 21–2
  component, 172
  monopolistic, 168
  neoclassical theory, 157
  pure and perfect, 67
  system, 172
complex competencies, 212
component competition, 172
computer industry, 167–82
  empirical analysis, 173–8
  industrial dynamics of, 167–70
  innovation and profit, 170–1, 174–6
  mass production, 169
  patentability of software, 261
  PDP8 minicomputer, 168–9
  R&D, 169
  standardization, 169
  workstations, 169
Conference Board/Corporate-Investor
    Summit, 54
consolidation, 153
consumer goods, 74
consumption, 39
contestable markets theory, 76
cooperation and assistance, 131, 137

coordination, 155, 156
corporate coherence, 144–5
corporate entrepreneurship, 6, 132,
    142, 144
  governance based on, 143–5
corporate governance, 30
corporate strategy, globalization of,
    250–1
cost shifting, 122
creative accumulation, 89
creative destruction, 23, 70, 89
credit rationing, 40, 225
credit risk assessment, 227
  qualitative analysis, 236–9
credit system, 113
crisis, 22–5
critical mass principle, 139
CSIRO, 99, 101

Data General, 168
decision trees, 140
default probability, 232
demand, 189–90
depression, 24
Dertouzos, Michael, 56
design, 72
development, 91
  and intellectual property rights,
    119–21
  role of finance in, 116–19
Digital, 168, 169
diversity, 170
Domar, Evsey, 14
dynamic factors, 156

Ebbers, Bernard, 51
economic change, 149–66
economic cycles, 19, 24, 27
economic efficiency, 121
economic growth, 72–5
Economic and Monetary Union,
    Stability and Growth Pact, 44
e-design, 91
e-engineering, 91
Eisinger, Jesse, 55
elementary competencies, 212
Elkind, Peter, *The Smartest Guys in the
    Room*, 51
Enron, 16, 51–3, 55

entrepreneurial alertness, 158, 186
entrepreneurial capital, 194
entrepreneurial discovery, 158
entrepreneurial firms, 160–3
entrepreneurial resources, 186–7
entrepreneurs, 5, 66, 67, 69, 149–66
  Anglo-Saxon tradition, 149
  Austrian tradition, 149, 157–60
  eviction of, 154–7
  heroic, 69–72
  Marshallian tradition, 149–50
  organic, 151–4
  place of, 83
  resource potential, 79–81
  socialized, 75–83
  see also innovation
environmental issues, 98–9
Eurodesk, 243
exogenous factors, 22
expectations, 162
exploitation, 184–7
  standard option, 185
exploration, 184–7
external flexibility, 250–1
external funding, 224–47

failure prediction, 228–9
Fairchild Semiconductor, 16
fiat money, 113
finance, x, 110–27
  and development, 116–19
  and technological innovation, 116–19
  transnational, 117
finance-based economy, 2, 3–6
financial capacity, 26, 210–12
financial capital, 112–16
financial competencies, 212, 214, 217–18, 219
financial innovations, xi, 30
financial markets, 188–91
  failure of, 97, 99
  liberalization of, 250–1
  see also market
financial reporting, 238
firms
  as 'black boxes', 248, 254
  co-evolution, 205

  creation of, 193–5
  entrepreneurial, 160–3
  exit from industry, 195
  heterogeneous, 198
  homogeneous, 196, 198
  innovative, 94, 142–6
  as institutions, 154–7
  joint-stock, 29
  mature, 136–7, 138
  splitting, 195
  start-up, 134–6, 138
  see also big firms; small and medium-sized enterprises
first mover advantage, 96
fiscal policy, 46, 47–8
fixed costs, 208–9
flexibility, 1, 32, 172, 250
  external, 250–1
  internal, 250–1
Ford Motors, 72, 75

Galbraith, John Kenneth, 14–15, 25, 72, 248
  Affluent Society, 265
  The New Industrial State, 14
Gas Bank, 52
Gates, Bill, 16
generally accepted accounting principles, 52
General Motors, 75
genetically modified organisms, 19
Gibbon, Edward, *The Decline and Fall of the Roman Empire*, 53
global capitalism, ix
global exploitation of technology, 103
global generation of technology, 102
globalization, 1
  and profitability imperative, 250–3
  of R&D, 356–7
  transnational capitalism, 110
global monetary standards, 114
global technological collaboration, 103
governance
  based on corporate entrepreneurship, 143–5
  innovative firms, 142–6

mature firms, 136–7, 138
start-up firms, 134–6, 138
gross domestic expenditures, 255
growth, 13, 66
  post-war economic, 72–5

Harvey, Campbell, 56
heroic entrepreneurs, 69–72
Hewlett-Packard, 168, 169
Home Depot, 55
homo economicus, 79
human capital, 194, 198

IBM, 75, 168
incrementalism, 92, 93, 95, 102
individualist professionals, xiv
industrial organization, 176–8
  and innovation, 171–2
  and profit, 172–3, 177
industrial property rights, 262–3
  see also intellectual property rights; patents
industrial revolution, 28, 69
industry, 188–91
  co-evolution of structure, 203
  global evolution, 196–9
industry life cycle, 131–48
  cooperative governance of start-ups, 134–6
  early stages, 133–4
  late stages, 134, 136–7
industry relative accounting ratio, 229
inflation barrier, 43–4
inflation gap, 38
information potential, 26
information technology, 76, 261
innovation, 2, 3, 5, 13–18, 35–49, 88, 133
  banking programmes, 240
  and capitalist efficiency, 121–3
  closed systems, 90
  and company size, 210–12
  computer industry, 170–1
  definition of, 21
  economies of scale, 209–10
  external funding, 224–47
  financial, xi, 30
  foundations of, 20–2
  and industrial organization, 171–2
  and institutional changes, 25–7
  and intellectual property rights, 119–21
  and long-lasting accumulation movements, 25–33
  and macroeconomics, 3–4, 8
  and networking, 257
  and profit, 6–9, 170–1, 174–6
  R&D in, 90–2
  radical, 22, 24, 96
  reorientation of, 123–4
  role of finance in, 116–19
  see also entrepreneurs
Innovation-Buy, 244
innovative behaviours, 138–42
innovative firms, 94
  governance of, 142–6
institutional changes, and innovation, 25–7
institutional investors, 30
institutions, rise of, 154–7
intangible investments, 93, 226–7
intangibles, 226
integration, 1
Intel, 15, 16
intellectual enclosures, 120
intellectual property rights, 99, 110–17, 136, 258–60
  and development, 119–21
  and innovation, 119–21
  patent pooling, 262–3
  protection of, 119–21
  see also patents
interest rates, 45
interest rate 'smoothing', 38
Intergraph, 169
internal flexibility, 250–1
internal rating systems, 229–39
  architecture, 232–6
  numerical, 235
  qualitative factors in, 236–9, 241
  scoring models, 234
international finance, 114–15
International Monetary Fund, 28
international political economy, 114
Internet, 19
Intesa Group, 242–3
IntesaNova, 242–3

investment, 35–49, 210–12
   and aggregate demand, 43–4
   intangible, 93
   investors, 135–6, 143–4
   diversity of, 251

Japan, 15
job seekers, 81, 82
Johnson, Carrie, 54
joint-stock firms, 29

Kaldor, Nicholas, 14
Kaleckian analysis, 41–4
Karpik, L., 20
Keynes, John Maynard, 58
Kirzner, I., 151, 157
Knight, Frank H., 150, 151–4
knowledge, 2, 31
   commercialization of, 117
   mutual, 158
   tacit, 260
knowledge-based economy, 2, 3–6
knowledge capital, 9, 252
   appropriation of, 258–63
   definition, 253
   formation of, 254–6
   protection of, 258–60
   role, 253–4
   socialization of, 256–8
knowledge creation, 143
knowledge economy, 26, 78–83
Koenig, Mark, 55
Kondratiev, N., 24
Kozlowski, Dennis, 16

labour market reform, 43
labour productivity, 122
Lafargue, Paul, 29
law of large numbers, 153
law of value, 112
Lay, Kenneth, 16, 51, 52
learning by doing, 36
least cost rule, 60
Lester, Richard, 56
leveraged buyout, 76
liberalization, 32, 76, 102
Likert scale, 173

liquidity, 211
loan pricing, 229
long-lasting accumulation
   movements, 25–33
   and social change, 27–33
loss given default, 232

McLean, Bethany, *The Smartest Guys in the Room*, 51
macroeconomics, 35–49
   and innovation, 3–4, 8
   new consensus, 37–41
management quality, 238
managerial capitalism, xiii, 72
managers, 72–5, 135–6
   option chains, 185–7
   rising power of, 249
market process, 157–60
   evolution of, 159
market share, 177
market-to-market accounting, 52
Marshall, Alfred, 149–51
Marx, Karl, 28, 71, 111, 248, 255
   Capital, 258
   Grundrisse, 264
mass consumption, 74
mass production, 169
mass unemployment, 77
mature firms, 136–7, 138
maximum productivity, 191–2
mergers and acquisitions, 115
Michelin, 259
Microsoft, 15, 16
monetary policy, 47–8
monopolistic competition, 168
monopoly capitalism, ix, 29
monopoly contestability, 1, 32
mutual knowledge, 158

NAIRU, 37, 43
Nash equilibrium, 141
National Science Board, 92
national security, 98
Nelson, Richard, 14
neoclassical theory
   competition, 157
   firms as 'black boxes', 248, 254
   growth, 13

networks, 26, 257
  links with innovation, 257
  network economies, 32
  networked firms, 78, 250–1, 257
    characteristics of, 78
new consensus in macroeconomics, 37–41
new product development, 90
non-accelerating inflation rate of unemployment, *see* NAIRU
non-productive behaviours, 50, 59
no-Ponzi game condition, 39
Noyce, Robert, 15–16

'onconmouse patent', 261
open innovation system, 90
opportunities, identification of, 185–6
options
  creation of, 187–95
  potential, 192–3
  real, 192–3, 197
option chains, 185–7
'organic solidarity', 251
organization of industry, 14, 134, 149–66
output gap, 38
outsourcing, 257
owners' capitalism, xiii
ownership structure, 238

Papandreou, Andreas, 14
Pareto efficiency, 121
Pareto optimality, 121
patent pools, 262–3
patents, 117, 256, 258–60, 262
  living organisms, 261
  scientific and technical, 261
  shortcomings of, 257–8
  see also intellectual property rights
pay-off ranking, 140
PDP8 minicomputer, 168–9
Penrose, Edith T., 151, 160–3
  The Theory of the Growth of the Firm, 160
physical capital, 194, 198
Pitt, Harvey, 54

Porter, Michael, 51, 62
  'Capital Disadvantage: America's Failing Capital Investment System', 57
  The Competitive Advantage of Nations, 56
post-industrial capitalism, ix
post-war economic growth, 72–5
potential options, 192–3
power
  of finance, 15
  of managers, 249
  market, 21
  monopoly, 95
  transfer to technostructure, 248–9
predation, 3, 13–18
predator constraints, 138–42
private equity firms, xi–xii
private financing, 5–6
private property, 73, 112, 124
private spending, 40
privatization, 1
produce differentiation, 169
production, 32
  organization of, 26
  socialization of, 118
production base, 163
productive labour, 122
productive opportunity, 161–2
productivity, 121
  labour, 122
  maximum, 191–2
  R&D, 209–10
professionals' capitalism, ix–x, xiii
profit, 26
  and industrial organization, 172–3, 177
  and innovation, 6–9, 170–1, 174–6
  and R&D intensity, 175
profitability imperative, 248–69
  and globalization, 250–3
profit lacuna, 50
property rights, *see* intellectual property rights
proprietary-system intensity, 176
prosperity, 24
pseudo-capitalists, 60–1
public policy, 77

qualitative analysis of credit risk, 236–9

R&D, 5, 88–109
  competence, 217, 218
  computer industry, 169, 173–4
  determinants of, 178
  external funding, 224–47
  fixed costs, 208–9
  global financial system, 101–4
  globalization of, 256–7
  informal, 91
  in innovation, 90–2
  institutionalization of, 89
  investment and financing, 93–7
  investment in, 190–1
  productivity, 209–10
  public policies, 97–101
radical innovation, 22, 24, 96
Rappaport, Alfred, 53
RCA, 95–6
Reagan–Volcker policies, x, 15
real options, 192–3, 197
recession, 24
recovery, 24
relationship banking, 226
renewal, 22–5
rentiers, xii–xiii, 58
research, 31, 90–1
  indivisibility of, 208–9
research and development, *see* R&D
resource potential, 79–81
resources, 161
Revisiting Stock Market Short-Termism, 54
rewards, 42
Richardson, G.B., 159
Ridge Computer, 169
Rigas, John, 51
risk, 163, 238
robber barons, xiv
Robinson, Joan, 58, 66, 248

sale of innovation competencies, 220, 221
Salter, W.E., 14
same-store comps, 55
Sanpaolo IMI Group, 243–4

Sarbanes-Oxley Corporate Antifraud Act (2002), 51, 57
Say, J.B., 254–5
Say's Law, 35, 58, 59
Schumpeter, Joseph, 19, 88–9, 157, 248
  Capitalism, Socialism and Democracy, 13
  Theory of Economic Evolution, 70
science, 20
  as capital asset, 21
Second World War, 66, 72, 254
secrecy, 258–60
services, 161
shareholders, 72–5
short-termism, xi, xiii–xiv, 4, 50–65
  definition, 51
  evidence for, 51–6
  history, 56–8
Silicon Graphics, 169
Skilling, Jeffrey, 51, 52, 55, 63
small and medium-sized enterprises, 207
  financial means, 210
  R&D costs, 211
  technological intensity, 215
Smith, Adam, 254
  Wealth of Nations, 258
social change, 27–33
socialization
  of capital, 51
  of knowledge capital, 256–8
  of production, 118
socialized entrepreneur, 75–83
social relations network, 79
social restructuring, 123–4
software patentability, 261
Solow, Robert, 56
specialization, 153
speculation, 28–9
SPRU database, 171
standardization, 169
start-up firms, 134–6, 138
stock exchange, 72
stock market churning, 57
Sullivan, Scott, 51
Sun Microsystems, 169
supply, 23, 189–90
supply and demand, 35

supply-side equilibrium, 43, 44, 47
survival, 184
system competition, 172

tacit knowledge, 260
Tandy, 169
Taylorist movement, 111
techno-globalism, 103
technological change, 26–7
technological intensity, 215
technology, 2
  co-evolution, 203–4
  global exploitation of, 103
  global generation of, 102
  maximum productivity, 191–2
technology alternatives, 191–2, 201–2
technology and market regimes, 199–202, 204
  market characteristics, 202
  technological characteristics, 200–2
technology transfer, 117, 257
technostructure, 66, 73, 89, 94–5
  transfer of power to, 248–9
Tobin Tax, 57
totalitarian capitalism, 110
Trade Related Industrial Property Rights, *see* TRIPs Agreement
trade secrets, 259
transaction costs, 136
transnational capitalism, 110
transnational finance, 117
TRIPs Agreement, 119, 120, 258, 261–2

uncertainty, 163
unemployment
  NAIRU, 37
  natural rate of, 38
Unicredit Group, 244
United States
  Antitrust Guidelines for the Licensing of Intellectual Property, 263
  Bayh Dole Act, 257
  corporate decline, 15
  Great Depression, 58
  National Science Board, 92
  short-termism, 50–65
universities
  as competitors, 264
  relations with industry, 257

value creation, 255
value theory, 114
venture capital, 134–5
vertical integration, 173
von Mises, L., 159

Walrasian paradigm, 7
White, Ben, 54
Winter, Sidney, 14
workstations, 169
World Bank, 77
WorldCom, 16, 51
World Trade Organization, 28, 262

zero-sum environment, 57